AF365884

A Guide to Research in

Electro Biochemical Technology

A Guide to Research in

Electro Biochemical Technology

Sevas Publishing
Sevas Educational Society
Andhra Pradesh, India
http://www.sevas.co.in

A Guide to Research Electro Biochemical Technology

Editor

R R Siva Kiran

Dept. of Biotechnology, M S Ramaiah Institute of Technology, MSR Nagar, Bangalore-560054, India (editor@jbiochemtech.com)

Associate Editor for special issue

Libuse Trnkova

Department of Chemistry, Faculty of Science, Masaryk University, Kotlarska 2, CZ-611 37 Brno, Czech Republic (libuse@chemi.muni.cz)

Special Issue Editorial Board

Rene Kizek

Department of Chemistry and Biochemistry, Faculty of Agronomy, Mendel University in Brno, Zemedelska 1, CZ-613 00 Brno, Czech Republic (kizek@sci.muni.cz)

Jaromir Hubalek

Department of Microelectronics, Faculty of Electrical Engineering and Communication, Brno University of Technology, Technicka 10, CZ-616 00 Brno, Czech Republic (hubalek@feec.vutbr.cz)

Vojtech Adam

Department of Chemistry and Biochemistry, Faculty of Agronomy, Mendel University in Brno, Zemedelska 1, CZ-613 00 Brno, Czech Republic (vojtech.adam@mendelu.cz)

Sevas Publishing Staff

R N Lakshmi Naidu, Publishing Editor; S Bindu Sowjanya Sree, Managing Editor;

Sidda Lingeswara, Publishing Staff

Scope

"Journal of Biochemical Technology" is a quarterly journal published by "Sevas Educational Society". The journal publishes advances in the field of biochemical research, biochemical engineering & bioinformatics. It provides a medium for the rapid publication of full-length articles, mini-reviews of new and emerging products and short communications on all aspects of biochemical technology. It is a unique source for scientists interested in both engineering as well as basic biological research.

Cover

Coverage includes enzymes and proteins; applied genetics and molecular biotechnology; computational biology, genomics and proteomics; metabolic & tissue engineering; medical, environmental, food, nano, electro and agro biotechnology.

Major Keywords:

Endo/exocytosis, Trafficking, Membrane Biology, Cell Migration, Cell-Matrix Organelle Biogenesis, Cytoskeleton Proteolysis, Cell Death, Cell Cycle, Cancer, Cell Growth/Death, Differentiation, Drug Targets, Gene Therapy, Models of Disease, Proteomics, Stem Cells, Bioenergetics, Mitochondria, Free Radicals, Redox Signaling, Ion Transport/Channels, Oxidative Stress, Cellular Metabolism, Xenobiotic Metabolism, Metabolomics, Bioenergetics, Mitochondria, Free Radicals, Redox Signalling Ion, Transport/ Channels, Oxidative Stress, Circadian Controls, Hormones, Photosynthesis, Signalling, Nitrogen fixation, Plant Disease, Transport, Plastid Systems, Cell Wall Synthesis, Phosphorylation, Protein kinases, Phosphatases, Ubiquitination, Lipid Signaling, Second Messengers, Cytokines, Receptors, Growth factors, G-Proteins, Integrins, Nitric Oxide, Enzymology, Molecular Mechanisms, Biophysics, Biomolecular Associations, Systems Biology, Chemical Biology, Structure, Proteins, Nucleic Acids, Amino acids, System Biology, Drug Design, Drug Targets, Biophysics, Chemical Biology, Biomolecules, Molecular Mechanisms, Enzymology, Lipids & Carbohydrates,E nviromental Biotechnology, Metabolic Engineering, Tissue Engineering, Nano biotechnology, Electro biotechnology, Agro biotechology & Medical biotechology.

Sevas Publishing

"Sevas Publishing" is maintained by a non-governmental, not-for-profit organization "Sevas Educational Society". "Seva" means selfless service in Sanskrit (An Ancient Indian Language), "The Society" was established on August 4th, 2006 at Komatipalli, a remote village near Visakhapatnam, South India. The society aims at the development of villages by educating people in various fields like biotechnology and bioinformatics. The society concentrates on distributing technologies for producing cost effective biofertilizers, bioinsecticides, single cell protein and biodiesel. The funds generated from this special issue of the journal will be directed towards village development.

Sevas Educational Society Council

A Guide to Research in Electro Biochemical Technology

Contents   *Page No.*

J Biochem Tech (2010) 2(5)
ISSN: 0974-2328

J Biochem Tech (2010) 2(5)
ISSN: 0974-2328

A Guide to Research in Electro Biochemical Technology

PREFACE

A special issue on "Physics and electro biochemical technology"

Physics, maths and chemistry have long played a dominant role in electronics, electrical, and other allied engineering sciences. However, wholesale implementation of life science concepts in electro-chemistry or nano-technology is relatively recent.

This special issue, entitled "A guide to research in electro-biochemical technology" aims at documenting state-of-the-art research, new developments, and directions for future investigation in the field of bio-physics, electro-biochemistry and nano-biotechnology. The book will guide the electrical, electronics and allied engineering branches to initiate the biology or biotechnology projects. The book also contains basic information about establishing electro-biochemical laboratory.

This special issue is an outgrowth of the *10th Workshop of Physical Chemists and Electrochemists*, which was held in Mendel University in Brno, Czech Republic. The workshop was dedicated to gain new knowledge of the physical, biophysical chemistry, electrochemistry and materials chemistry, which were enriched by the initial experience from nano-science and nano-materials.

The book is divided into three sections. First section is dedicated for understanding the terminology used in the research articles (This section will help biologist to understand electro-terminology and electrochemists to understand bio-terminology). The next section contains 57 peer-reviewed manuscripts on the topics of spectroscopy, electrochemistry, nanobiotechnology, chromatography, environmental studies and protein & nucleotides. The last section contains information about the list of equipments required for executing electro-biochemical research projects. We have also included detailed specifications, approximate costs and top manufacturers in the world. The book comes with a full package of media resources which contains basic biology concepts, free links to animations, top manufacturers, local vendors, equipment websites and extra facilities required for efficient execution of electro-bio projects.

The media resources can be accessed via link http://www.sevas.us/1/electrobiochem.html (Username: xscfg; Password: jkildh; Purchase details)

Who can use this book?

Electrical, electronics and allied sciences
This book will be helpful for electrical, electronics and allied engineering science departments to initiate biology/biotec.hnology related interdisciplinary research projects. The book also gives a brief idea about up gradation of available facilities.

Biology, biotechnology and life sciences
This book will also be helpful for biology, biotechnology and life science departments to initiate electro or nano related interdisciplinary research.

Researchers
We have included fifty seven research works for the benefit of students and faculties. Small projects or old projects were also included for helping students in developing countries.

How to read this book?

We suggest reading the last section for understanding the basic facilities required to carry out electro-biochemical projects. You can also take help of glossary section or media resources, if you face difficulty in understanding the manuscripts.

How to establish electro biochemical lab?

Electro-biochemical lab involves the combination of bio equipments and electro-chemical equipments. For example Thermal Cycler, Micro centrifuge, UV/Visible spectrometer and pH/conductivity meter are basic equipments present in all bio labs and Potentiostat or Galvanostat are present in electrical labs. Collaboration with biology or life science or electronics/electrical or allied engineering departments is required before procurement of any equipment. The last section and the media resources will help both departments to understand the base facilities required to initiate electro-biochemical research.

Edited by R R Siva Kiran and Libuse Trnkova

J Biochem Tech (2010) 2(5): ii-ix
ISSN: 0974-2328

Glossary

The following glossary is intended to provide a general understanding of some terms used by biologists and electrochemists. The list focuses on some of the important aspects for understanding the research articles published in this special issue.

Acute toxicity: Acute toxicity describes the adverse effects of a substance that result either from a single exposure or from multiple exposures in a short space of time (usually less than 24 hours). To be described as acute toxicity, the adverse effects should occur within 14 days of the administration of the substance.

Amperometry: Amperometry in chemistry and biochemistry is detection of ions in a solution based on electric current or changes in electric current. During amperometric assays the potential of the indicator electrode is adjusted to a value on the plateau of the voltammetric wave. The current that flows between the indicator electrode and a second electrode in the solution is measured and related to the concentration of the analyte.

Antistatic Agent: An antistatic agent is a compound used for treatment of materials or their surfaces in order to reduce or eliminate buildup of static electricity generally caused by the triboelectric effect. Its role is to make the surface or the material itself slightly conductive, either by being conductive itself, or by absorbing moisture from the air, so some humectants can be used. The molecules of an antistatic agent often have both hydrophilic and hydrophobic areas, similar to those of a surfactant; the hydrophobic side interacts with the surface of the material, while the hydrophilic side interacts with the air moisture and binds the water molecules.

Anodization: Anodizing, or anodising is an electrolytic passivation process used to increase the thickness of the natural oxide layer on the surface of metal parts. The process is called "anodizing" because the part to be treated forms the anode electrode of an electrical circuit. Anodizing increases corrosion resistance and wear resistance, and provides better adhesion for paint primers and glues than does bare metal. Anodic films can also be used for a number of cosmetic effects, either with thick porous coatings that can absorb dyes or with thin transparent coatings that add interference effects to reflected light. Anodizing is also used to prevent galling of threaded components and to make dielectric films for electrolytic capacitors. Anodic films are most commonly applied to protect aluminium alloys, although processes also exist for titanium, zinc, magnesium, niobium, and tantalum.

Biosensor: A biosensor is an analytical device for the detection of an analyte that combines a biological component with a physicochemical detector component.

It consists of 3 parts:

- The sensitive biological element (biological material (e.g. tissue, microorganisms, organelles, cell receptors, enzymes, antibodies, nucleic acids, etc.), a biologically derived material or biomimic component that interacts (binds or recognises) the analyte under study. The biologically sensitive elements can also be created by biological engineering.
- The transducer or the detector element (works in a physicochemical way; optical, piezoelectric, electrochemical, etc.) that transforms the signal resulting from the interaction of the analyte with the biological element into another signal (i.e., transducers) that can be more easily measured and quantified;
- Biosensor reader device with the associated electronics or signal processors that are primarily responsible for the display of the results in a user-friendly way. This sometimes accounts for the most expensive part of the sensor device, however it is possible to generate a user friendly display that includes transducer and sensitive element. The readers are usually custom-designed and manufactured to suit the different working principles of biosensors.

Cancerogenic: Carcinogen or cancerogenic or carcinogenic is any substance which causes cancer.

Cell lines: A cell line is a permanently established cell culture that will proliferate indefinitely given appropriate fresh medium and space. Cell culture is the complex process by which cells are grown under controlled conditions. In practice, the term "cell culture" has come to refer to the culturing of cells derived from single cellular eukaryotes, especially animal cells. The establishment of cell lines from human tissue is difficult.

Chelation: Chelation is the formation or presence of two or more separate coordinate bonds between a polydentate (multiple bonded) ligand and a single central atom.

Chemotherapy: Chemotherapy is the treatment of cancer with an antineoplastic drug (any of several drugs that control or kill neoplastic cells or tumour cells) or with a combination of such drugs into a standardized treatment regimen. Most commonly, chemotherapy acts by killing cells that divide rapidly, one of the main properties of most cancer cells. This means that it also harms cells that divide rapidly under normal circumstances: cells in the bone marrow, digestive tract and hair follicles. This results in the most common side effects of chemotherapy: myelosuppression (decreased production of blood cells, hence also immunosuppression), mucositis (inflammation of the lining of the digestive tract), and alopecia (hair loss).

Chromatography: Chromatography, literally "color writing", was used and named in the first decade of the 20th century, primarily for the separation of plant pigments such as chlorophyll, carotenes, and xanthophylls. It is the collective term for a set of laboratory techniques for the separation of mixtures. The mixture is dissolved in a fluid called the "mobile phase", which carries it through a structure holding another material called the "stationary phase". The various constituents of the mixture travel at different speeds, causing them to separate. The separation is based on differential partitioning between the mobile and stationary phases. Subtle differences in a compound's partition coefficient result in differential retention on the stationary phase and thus changing the separation. High-performance liquid chromatography (HPLC), is a chromatographic technique that can separate a mixture of compounds and is used in biochemistry and analytical chemistry to identify, quantify and purify the individual components of the mixture. HPLC typically utilizes different types of stationary phases, a pump that moves the mobile phase(s) and analyte through the column, and a detector to provide a characteristic retention time for the analyte. The detector may also provide additional information related to the analyte, (i.e. UV/Vis spectroscopic data for analyte if so equipped). Analyte retention time varies depending on the strength of its interactions with the stationary phase, the ratio/composition of solvent(s) used, and the flow rate of the mobile phase. It is a form of liquid chromatography that utilizes smaller column size, smaller media inside the column, and higher mobile phase pressures.

Chronopotentiometry: It is a form of electroanalysis based on the measurement at a working electrode of the rate of change in potential versus time; the current is controlled. Chronopotentiometry (CP) is the most basic constant current experiment. In CP, a current step is applied across an electrochemical cell (without stirring). In double step chronopotentiometry (DSCP), a second current step is applied. It should be noted that the time scale of a DSCP experiment is typically shorter (seconds or milliseconds) than that of CP experimm (minutes or seconds). The protocol for defining the sampling rate is therefore different for the two techniques. CP is a standard technique, whereas DSCP is only available as an optional addition.

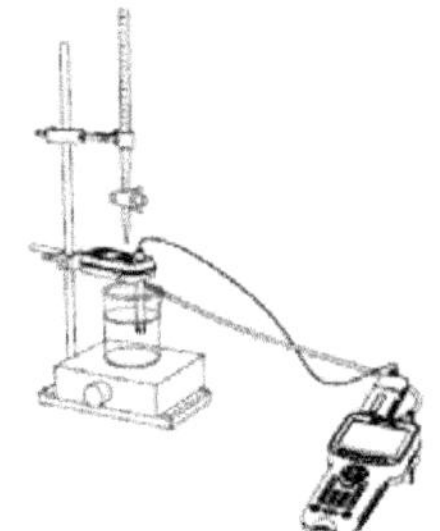

Conductometric titrations: Titration is the a method of determining the concentration of an unknown solution (the analyte) by reacting it completely with a standardized reagent that is a solution of known concentration (the titrant). The point at which all of the analyte is consumed is the equivalence point. The number of moles of analyte is calculated from the volume of reagent that is required to react with all of the analyte, the titrant concentration, and the reaction stoichiometry. A pH electrode was used to measure the pH of the analyte and to determine when the equivalence point was reached. Another method of determining when the equivalence point is reached is by using a conductivity sensor to measure the conductivity of the analyte. The conductivity of a solution is dependent on several factors, including the concentration of the solute, the degree of dissociation of the solute, the valence of the ion(s) present in the solution, the temperature, and the mobility of the ions in the solution.

Ref: http://www.vrml.k12.la.us/rpautz/documents/Chemistry/ConductometricTitration.pdf

Conductometry: Measurement of the electrical conductivity of a solution during the course of a chemical reaction. It is the determination of the quantity of a material present in a mixture by measurement of its effect on the electrical conductivity of the mixture.

Cyclic voltammetry: In a cyclic voltammetry experiment the working electrode potential is ramped linearly versus time like linear sweep voltammetry. Cyclic voltammetry takes the experiment a step further than linear sweep voltammetry which ends when it reaches a set potential. When cyclic voltammetry reaches a set potential, the working electrode's potential ramp is inverted. This inversion can happen multiple times during a single experiment. The current at the working electrode is plotted versus the applied voltage to give the cyclic voltammogram trace. Cyclic voltammetry is generally used to study the electrochemical properties of an analyte in solution.

Ref:http://www.mpip-mainz.mpg.de/~andrienk/journal_club/cyclic_voltammetry.pdf

Cytotoxicity: Cytotoxicity is the quality of being toxic to cells.

Differential pulse voltammetry: Differential Pulse Voltammetry is often used to make electrochemical measurements. It can be considered as a derivative of linear sweep voltammetry (LSV) or staircase voltammetry, with a series of regular voltage pulses superimposed on the potential linear sweep or stair steps. Differential pulse voltammetry adds a periodically applied potential pulse (temporary increase in potential) to the voltage ramp used for LSV. The current is measured just prior to application of the pulse and at the end of the applied pulse. The difference between the two currents is plotted as a function of the LSV ramp potential. Pulse voltammetry utilizes a regularly increasing pulse height.

DNA: Deoxyribonucleic acid (DNA) is a nucleic acid that contains the genetic instructions used in the development and functioning of all known living organisms.

DNA adduct: A DNA adduct is a piece of DNA covalently bonded to a (cancer-causing) chemical.

EC50: The term half maximal effective concentration (EC50) refers to the concentration of a drug, antibody or toxicant which induces a response halfway between the baseline and maximum after some specified exposure time. It is commonly used as a measure of drug's potency. The EC50 of a graded dose response curve therefore represents the concentration of a compound where 50% of its maximal effect is observed. The EC50 of a quantal dose response curve represents the concentration of a compound where 50% of the population exhibit a response, after a specified exposure duration.

Electrocardiography: Electrocardiography (ECG or EKG from the German Elektrokardiogramm) is a transthoracic (across the thorax or chest) interpretation of the electrical activity of the heart over a period of time, as detected by electrodes attached to the outer surface of the skin and recorded by a device external to the body. The recording produced by this noninvasive procedure is termed an electrocardiogram (also ECG or EKG).

Electrochemical deposition: Electrodeposition or electrochemical deposition (of metals or alloys) involves the reduction of metal ions from aqueous, organic, or fused salt electrolytes. In it's simplest form the reaction in aqueous medium at the cathode follows the equation (1) with a corresponding anodic reaction. The anode material can either be the metal to be deposited (in this case the electrode reaction is electrodissolution that continuously supplies the metal ions) or the anode can be an inert material and the anodic reaction is oxygen evolution (in this case the plating solution is eventually depleted of metal ions).

$$M+n + ne- ==> M \qquad (1)$$

The deposition may, in principle, be accomplished via two different paths:

- An electrodeposition process in which electrons are provided by an external power supply.
- An " electroless (autocatalytic) deposition process in which a reducing agent in solution is the electron source.

The deposition reaction presented in Equation [1] is a reaction of charged particles at the interface between a solid (metal) electrode and a liquid solution. The two types of charged particles that can cross the interface are metal ions "M+n" and electrons "e-".

Ref: http://electrochem.cwru.edu/encycl/art-e01-electroplat.htm
http://cnx.org/content/m22370/latest/

Electrochemical polymerization: Polymerization is a process of reacting monomer molecules together in a chemical reaction to form three-dimensional networks or polymer chains. In an electrochemical polymerization, a potential is applied across a solution containing the compound of interest and a suitable electrolyte, producing a conductive film on the anode. Electrochemical polymerization is convenient, since the polymer does not need to be isolated and purified, but it can produce polymers with undesirable alpha-beta linkages and varying degrees of regioregularity.

Ref: http://www.mjcce.org.mk/PDF/25_1_129.pdf

Electrolyte: An electrolyte is any substance containing free ions that make the substance electrically conductive. The most typical electrolyte is an ionic solution, but molten electrolytes and solid electrolytes are also possible.

Electrophoretic analysis: Running electric currents through tissues in a chemical medium or gel to cause various components to separate. Used in genetic analysis.

Electrophysiological studies: An electrophysiology study (EP test or EP study) is a minimally invasive procedure which tests the electrical conduction system of the heart to assess the electrical activity and conduction pathways of the heart. The study is indicated to investigate the cause, location of origin, and best treatment for various abnormal heart rhythms. This type of study is performed by a specialist (an electrophysiologist) and is performed using a single or multiple catheters situated within the heart through a vein or artery.

Electroplating: Electroplating is a plating process in which metal ions in a solution are moved by an electric field to coat an electrode. The process uses electrical current to reduce cations of a desired material from a solution and coat a conductive object with a thin layer of the material, such as a metal. Electroplating is primarily used for depositing a layer of material to bestow a desired property (e.g., abrasion and wear resistance, corrosion protection, lubricity, aesthetic qualities, etc.) to a surface that otherwise lacks that property. Another application uses electroplating to build up thickness on undersized parts.

Emery paper: Emery paper is a type of paper that can be used for sanding down hard and rough surfaces. It can also be used for resistant technology purposes to give a smooth, shiny finish to manufactured products and is often used in the finishing of high-end watch movements.

EPR experiments: Electron paramagnetic resonance (EPR) or electron spin resonance (ESR) spectroscopy is a technique for studying chemical species that have one or more unpaired electrons, such as organic and inorganic free radicals or inorganic complexes possessing a transition metal ion. The basic physical concepts of EPR are analogous to those of nuclear magnetic resonance (NMR), but it is electron spins that are excited instead of spins of atomic nuclei. Because most stable molecules have all their electrons paired, the EPR technique is less widely used than NMR. However, this limitation to paramagnetic species also means that the EPR technique is one of great specificity, since ordinary chemical solvents and matrices do not give rise to EPR spectra.

http://pet.radiology.uiowa.edu/downloads/Xiang%20Wu%20Papers/Core%20papers/SFRS2002Kuppusamy-epr.pdf

Enzymatic activity: Enzymes are proteins that catalyze (i.e., increase the rates of) chemical reactions. In enzymatic reactions, the molecules at the beginning of the process, called substrates, are converted into different molecules, called products. Almost all chemical reactions in a biological cell need enzymes in order to occur at rates sufficient for life. Since enzymes are selective for their substrates and speed up only a few reactions from among many possibilities, the set of enzymes made in a cell determines which metabolic pathways occur in that cell. Enzyme assays are laboratory methods for measuring enzymatic activity. They are vital for the study of enzyme kinetics and enzyme inhibition. Enzyme activity = moles of substrate converted per unit time = rate × reaction volume. Enzyme activity is a measure of the quantity of active enzyme present and is thus dependent on conditions, which should be specified.

Fluorescence spectroscopy: It is a type of electromagnetic spectroscopy which analyzes fluorescence from a sample. It involves using a beam of light, usually ultraviolet light, that excites the electrons in molecules of certain compounds and causes them to emit light of a lower energy, typically, but not necessarily, visible light. (http://www.bioch.ox.ac.uk/aspsite/services/equipmentbooking/biophysics/introfluor.pdf)

Glossy paper: Glossy photo quality paper is designed to make printed photos look sharp, vibrant and more like a traditional photograph. Handling of glossy paper is important as the paper can quickly and easily be marred by fingerprints and dirt.

Grams32 software: The complete spectroscopy software suite for all your data importing, processing, viewing, organizing, and accessing needs. Approximate Price: USD 3500/- (Developed by Galactic Industries Corporation and supplied by Thermo Scientific Ltd. http://www.thermoscientific.com).

HL–60 myeloid cell line: The HL60 cell line originated from a female patient with acute myeloid leukaemia (i.e cancer).

For more information: http://www.ncbi.nlm.nih.gov/pmc/articles/PMC2149104/pdf/brjcancersuppl00067-0046.pdf

Hyperchromicity: Hyperchromicity is the increase of absorbance (optical density) of a material. The most famous example is a hyperchromicity of a DNA that occurs when DNA duplex is denatured. The UV absorption is increased when the two single DNA strands are being separated, either by heat or by addition of denaturant or by increasing the pH level. The opposite, a decrease of absorbance is called hypochromicity.

Humic substances: Humic acid is a principal component of humic substances, which are the major organic constituents of soil (humus), peat, coal, many upland streams, dystrophic lakes, and ocean water. It is produced by biodegradation of dead organic matter.

ICP–MS: Inductively coupled plasma mass spectrometry (ICP-MS) is a type of mass spectrometry that is highly sensitive and capable of the determination of a range of metals and several non-metals at concentrations below one part in 1012 (part per trillion). It is based on coupling together inductively coupled plasma as a method of producing ions (ionization) with a mass spectrometer as a method of separating and detecting the ions.

Infrared spectroscopy: Infrared spectroscopy (IR spectroscopy) is the spectroscopy that deals with the infrared region of the electromagnetic spectrum, that is light with a longer wavelength and lower frequency than visible light. It covers a range of techniques, mostly based on absorption spectroscopy. As with all spectroscopic techniques, it can be used to identify and study chemicals. A common laboratory instrument that uses this technique is a Fourier transform infrared (FTIR) spectrometer.

Immunology: Immunology is a broad branch of biomedical science that covers the study of all aspects of the immune system in all organisms. It deals with the physiological functioning of the immune system in states of both health and diseases; malfunctions of the immune system in immunological disorders (autoimmune diseases, hypersensitivities, immune deficiency, transplant rejection); the physical, chemical and physiological characteristics of the components of the immune system in vitro, in situ, and in vivo. Immunoseparation methods include affinity methods with antibodies bound to magnetic beads, biodegradable beads, non-biodegradable beads, to panning surfaces including dishes, and to combinations of these methods. Immunoprecipitation (IP) is the technique of precipitating a protein antigen out of solution using an antibody that specifically binds to that particular protein. This process can be used to isolate and concentrate a particular protein from a sample containing many thousands of different proteins. Immunoprecipitation requires that the antibody be coupled to a solid substrate at some point in the procedure.

Impedance or Electrical impedance: Electrical impedance, or simply impedance, describes a measure of opposition to alternating current (AC). Electrical impedance extends the concept of resistance to AC circuits, describing not only the relative amplitudes of the voltage and current, but also the relative phases. When the circuit is driven with direct current (DC), there is no distinction between impedance and resistance; the latter can be thought of as impedance with zero phase angle.

Isolation of DNA: The procedure is suitable for all types of tissues from wide variety of animal, blood and plant species. All DNA extraction steps are performed at weak acid pH (HEPES or MOPS free acids) and optionally with hot chloroform for 'difficult' samples, and at room temperature.

The following protocol is designed for small and large tissue samples (tissue volume 100-200 µl). Note that isolating genomic DNA not requires gentle mixing because the DNA not be sheared by vortexing.

CTAB (Cetyl trimethylammonium bromide) or (Guanidine thiocyanate) method for DNA extraction protocol

Materials for total DNA isolation

- CTAB solution: 1-2% CTAB (, 2 M NaCl, 10 mM Na3EDTA, 0.1 M MOPS (HEPES), pH ? 4.6;
  1-2 g CTAB, 2.1 g MOPS-acid, 2 ml 0.5 M Na3EDTA, 40 ml 5 M NaCl;

- GuTC extraction buffer: 1 M Guanidine thiocyanate, 10 mM Na3EDTA, 0.1 M MOPS, pH ? 4.6; the final concentration of guanidine thiocyanate may need to optimized for certain plant/animals tissue from 0.5-1-2 M, but high concentration of guanidine thiocyanate (>1 M) may negatively interfere with the provision of high quality DNA for tissues with high concentration of pectin or polysaccharides (recommend using a concentration of 0.5 M of guanidine thiocyanate for tissues with a high concentration of polysaccharides, both in plants and animals);

- Chloroform-isoamyl alcohol mix (24:1);
- 100% isopropanol (isopropyl alcohol, 2-propanol);
- 70% ethanol;
- 10 M lithium chloride;
- Fresh Milli-Q water (or Milli-Q ultrapure BioPak water) or 1xTE (1 mM EDTA, 10mM Tris-HCl, pH 8.0) or 1xTHE (1 mM EDTA, 5 mM Tris, 5 mM HEPES, pH 8.0). When an ultrafiltration cartridge (BioPak) is utilized at the point-of-use, the water is suitable for genomics applications (quality at least equivalent to DEPC-treated water) and cell culture.

DNA extraction protocol

1. 2 ml Eppendorf Safe-Lock microcentrifuge tube with tissue sample and glass boll freeze at -80°C, grind in the MM300 Mixer Mill for 3 min at 30 Hz.
2. In 2 ml tube with mechanically disrupted seeds or leaves or herbarium or blood or DNA solution (CTAB purification) add fresh 1 ml CTAB or GuTC solution buffer (the sample volume should not exceed 20% of lysis buffer), vortex very well; optional: add 0.5 ml of chloroform, vortex very well (in the MM300 Mixer Mill for 3 min at 30 Hz); incubate the samples at 65°C within an hour to overnight.
3. Spin at maximum speed in a microcentrifuge for 2 minutes, transferred the upper aqueous layer to a new 2 ml microcentrifuge tube.
4. Add 0.5 ml of chloroform, vortex very well for 2 minute creating an emulsion (in the MM300 Mixer Mill at 30 Hz).
5. Spin at maximum speed in a microcentrifuge for 5 minutes.
6. Transferred the upper aqueous layer to a new 2 ml microcentrifuge tube which contains of 0.8 ml 2-propanol, vortex well and centrifuge the tubes at maximum speed in a microcentrifuge for 3 minutes.
7. Discard supernatant and wash pellet by adding 1.8 ml 70% EtOH, vortex well. Centrifuge at 14,000 rpm for 2 min and discard ethanol.
8. The DNA/RNA pellet do not dry and dissolved immediately in 300 µl 1xTE, pH 8.0 (with RNAse A) at 55°C for 10-20 minutes.

Source: http://primerdigital.com/dna.html

LabSpec: LabSpec is a complete Raman software that takes you all the way from system set up, through data measurement to data processing and report generation. The software is supplied by Horiba Scientific Inc, Japan.

(http://www.horiba.com/scientific/products/raman-spectroscopy/software/labspec-functionality/).

Lanthanides: The lanthanide elements are the group of elements with atomic number increasing from 57 (lanthanum) to 71 (lutetium). They are termed lanthanide because the lighter elements in the series are chemically similar to lanthanum. (57. lanthanum (La), 58. cerium (Ce), 59. praseodymium (Pr), 60. neodymium (Nd), 61. promethium (Pm), 62. samarium (Sm), 63. europium (Eu), 64. gadolinium (Gd), 65. terbium (Tb), 66. dysprosium (Dy), 67. holmium (Ho), 68. erbium (Er), 69. thulium (Tm), 70. ytterbium (Yb), 71. lutetium (Lu)).

Laser Ablation Inductively Coupled Plasma Mass Spectrometry: In Laser Ablation Inductively Coupled Plasma Mass Spectrometry the sample is directly analyzed by ablating with a pulsed laser beam. The created aerosols are transported into the core of inductively coupled argon plasma (ICP), which generates temperature of approximately 8000°C. The plasma in ICP-MS is used to generate ions that are then introduced to the mass analyzer. These ions are then separated and collected according to their mass to charge ratios. The constituents of an unknown sample can then be identified and measured. ICP-MS offers extremely high sensitivity to a wide range of elements. For laser ablation, any type of solid sample can be ablated for analysis; there are no sample-size requirements and no sample preparation procedures. Chemical analysis using laser ablation requires a smaller amount of sample (micrograms) than that required for solution nebulization (milligrams). Depending on the analytical measurement system, very small amount of sample quantities may be sufficient for this technique. In addition, a focused laser beam permits spatial characterization of heterogeneity in solid samples, with typically micron resolution both in terms of lateral and depth conditions.

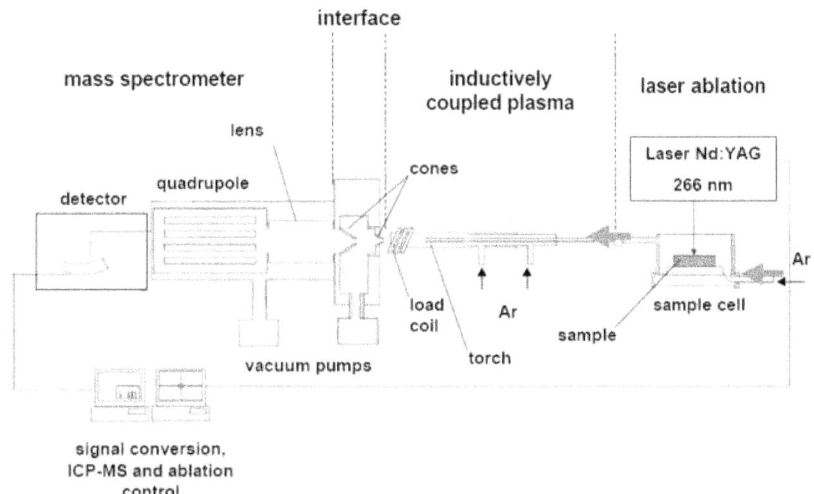

Systematic setup of Laser Ablation Inductively Coupled Plasma Mass Spectrometry (LA-ICP-MS)

Ref: http://www.si.edu/mci/downloads/reports/icp-dussubieux.pdf

Laser Induced Breakdown Spectroscopy: Laser-induced breakdown spectroscopy (LIBS) is a type of atomic emission spectroscopy which uses a highly energetic laser pulse as the excitation source. The laser is focused to form a plasma, which atomizes and excites samples. In principle, LIBS can analyse any matter regardless of its physical state, be it solid, liquid or gas. A typical LIBS system consists of a neodymium doped yttrium aluminium garnet (Nd:YAG) solid-state laser and a spectrometer with a wide spectral range and a high sensitivity, fast response rate, time gated detector. This is coupled to a computer which can rapidly process and interpret the acquired data. As such LIBS is one of the most experimentally simple spectroscopic analytical techniques, making it one of the cheapest to purchase and to operate.

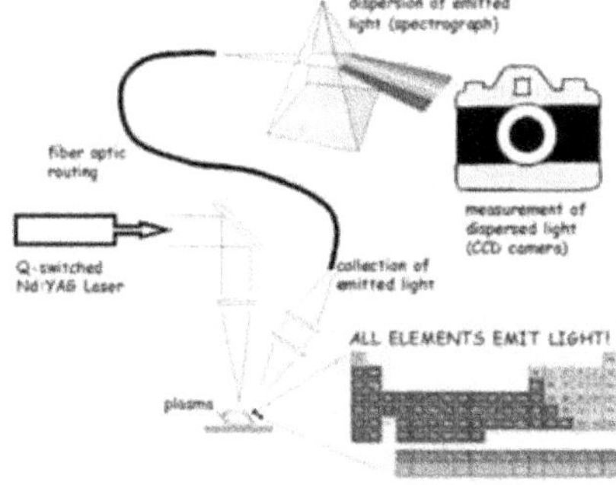

The Nd:YAG laser generates energy in the near infrared region of the electromagnetic spectrum, with a wavelength of 1064 nm. The pulse duration is in the region of 10 ns generating a power density which can exceed 1 GW·cm-2 at the focal point. Other lasers have been used for LIBS, mainly the Excimer (Excited dimer) type that generates energy in the visible and ultraviolet regions.

The spectrometer consists of either a monochromator (scanning) or a polychromator (non-scanning) and a photomultiplier or CCD detector respectively. The most common monochromator is the Czerny-Turner type whilst the most common polychromator is the Echelle type. However, even the Czerny-Turner type can be (and is often) used to disperse the radiation onto a CCD effectively making it a polychromator. The polychromator spectrometer is the type most commonly used in LIBS as it allows simultaneous acquisition of the entire wavelength range of interest.

The spectrometer collects electromagnetic radiation over the widest wavelength range possible, maximising the number of emission lines detected for each particular element. Spectrometer response is typically from 1100 nm (near infrared) to 170 nm (deep ultraviolet), the approximate response range of a CCD detector. All elements have emission lines within this wavelength range. The energy resolution of the spectrometer can also affect the quality of the LIBS measurement, since high resolution systems can separate spectral emission lines in close juxtaposition, reducing interference and increasing selectivity. This feature is particularly important in specimens which have a complex matrix, containing a large number of different elements. Accompanying the spectrometer and detector is a delay generator which accurately gates the detector's response time, allowing temporal resolution of the spectrum.

J Biochem Tech (2010) 2(5): ii-ix

LC50: In toxicology, the median lethal dose, LD50 (abbreviation for "Lethal Dose, 50%"), LC50 (Lethal Concentration, 50%) or LCt50 (Lethal Concentration & Time) of a toxin, radiation, or pathogen is the dose required to kill half the members of a tested population after a specified test duration. LD50 figures are frequently used as a general indicator of a substance's acute toxicity.

Ligands: A ligand is an ion or molecule (see also: functional group) that binds to a central metal atom to form a coordination complex. The bonding between metal and ligand generally involves formal donation of one or more of the ligand's electron pairs.

Linear sweep voltammetry: Linear sweep voltammetry is a voltammetric method where the current at a working electrode is measured while the potential between the working electrode and a reference electrode is swept linearly in time. Oxidation or reduction of species is registered as a peak or trough in the current signal at the potential at which the species begins to be oxidized or reduced.

Ref:http://www.mpip-mainz.mpg.de/~andrienk/journal_club/cyclic_voltammetry.pdf

Luria-Bertani medium: Luria-Bertani medium, a nutritionally rich medium, is primarily used for the growth of bacteria. These media have been widely used in molecular microbiology applications for the preparation of plasmid DNA and recombinant proteins.

Preparing Luria-Bertani medium (Makes 1000 mL)
- Measure about 700 mL deionized water into a graduated cylinder.
- Mix following into water: 10 g Bactotryptone, 5 g Bacto yeast, and 5 g NaCl.
- Use stir bar to mix well. While stirring, add 300 mL deionized water, making sure to wash sides of cylinder.
- Calibrate pH meter with 7.0 and 10.0 pH solutions. Never leave meter out of solution too long. Rinse meter with deionized water after dipping into and gently with Kimwipes.
- While broth mixes, adjust its pH to 7.5 with dropwise additions of 5 M NaOH and 1M HCl.
- Pour broth into a 1000 mL autoclavable bottle, and autoclave on a liquid cycle.
- Once broth reaches room temperature, store in cold room.

Ref: http://www.bcm.edu/physio/lab_pages/pedersen/protocols.html#Luria%20Bertani%20Medium

MALDI–TOF: Matrix-assisted laser desorption/ionization (MALDI) is a soft ionization technique used in mass spectrometry, allowing the analysis of biomolecules (biopolymers such as DNA, proteins, peptides and sugars) and large organic molecules (such as polymers, dendrimers and other macromolecules), which tend to be fragile and fragment when ionized by more conventional ionization methods. It is similar in character to electrospray ionization both in relative softness and the ions produced (although it causes many fewer multiply charged ions).

The MALDI is a two step process. First, desorption is triggered by a UV laser beam. Matrix material heavily absorbs UV laser light, leading to the ablation of upper layer (~micron) of the matrix material. A hot plume produced during the ablation contains many species : neutral and ionized matrix molecules, protonated and deprotonated matrix molecules, matrix clusters and nanodroplets. The second step is ionization (more accurately protonation or deprotonation). Protonation (deprotonation) of analyte molecules takes place in the hot plume. Some of the ablated species participate in protonation (deprotonation) of analyte molecules. The type of a mass spectrometer most widely used with MALDI is the TOF (time-of-flight mass spectrometer), mainly due to its large mass range. The TOF measurement procedure is also ideally suited to the MALDI ionization process since the pulsed laser takes individual 'shots' rather than working in continuous operation. MALDI-TOF instrument or reflectron is equipped with an "ion mirror" that reflects ions using an electric field, thereby doubling the ion flight path and increasing the resolution. Today, commercial reflectron TOF instruments reach a resolving power m/Δm of well above 20'000 FWHM (full-width half-maximum, Δm defined as the peak width at 50% of peak height).

Metabolites: Metabolites are the intermediates and products of metabolism. The term metabolite is usually restricted to small molecules. A primary metabolite is directly involved in normal growth, development, and reproduction. Alcohol is an example of a primary metabolite produced in large-scale by industrial microbiology. A secondary metabolite is not directly involved in those processes, but usually has an important ecological function. Examples include antibiotics and pigments.

Micrograph: A micrograph or photomicrograph is a photograph or similar image taken through a microscope or similar device to show a magnified image of an item. A microphotograph is a very small picture shown large like a real picture, e.g. a microdot. Microphotography is the production of such very small photographs. Two important applications of microphotography were the creation of computer chips and the creation of microdots to hide information.

Nanobiosensors: A biosensor is a measurement system for the detection of an analyte that combines a biological component with a physicochemical detector, and a nanobiosensor is a biosensor that on the nano-scale size.

Ref: http://frontpage.okstate.edu/nanotech/Reports/2007/Presentations/Cagri%20Ozge%20Topal.ppt

Nanostructure: A nanostructure is an object of intermediate size between molecular and microscopic (micrometer-sized) structures. In describing nanostructures it is necessary to differentiate between the number of dimensions on the nanoscale. Nanotextured surfaces have one dimension on the nanoscale, i.e., only the thickness of the surface of an object is between 0.1 and 100 nm. Nanotubes have two dimensions on the nanoscale, i.e., the diameter of the tube is between 0.1 and 100 nm; its length could be much greater. Finally, spherical nanoparticles have three dimensions on the nanoscale, i.e., the particle is between 0.1 and 100 nm in each spatial dimension. The terms nanoparticles and ultrafine particles (UFP) often are used synonymously although UFP can reach into the micrometre range. The term 'nanostructure' is often used when referring to magnetic technology.

Nanotubes: A nanotube is a nanometer-scale tube-like structure. Eg: Carbon nanotube, Inorganic nanotube, DNA nanotube, Membrane nanotube - a tubular membrane connection between cells. Carbon nanotubes (CNTs) are allotropes of carbon with a cylindrical nanostructure. Nanotubes have been constructed with length-to-diameter ratio of up to 132,000,000:1, significantly larger than for any other material. These cylindrical carbon molecules have unusual properties, which are valuable for nanotechnology, electronics, optics and other fields of materials science and technology. In particular, owing to their extraordinary thermal conductivity and mechanical and electrical properties, carbon nanotubes may find applications as additives to various structural materials.

NMR spectroscopy: Nuclear magnetic resonance spectroscopy, most commonly known as NMR spectroscopy, is a research technique that exploits the magnetic properties of certain atomic nuclei to determine physical and chemical properties of atoms or the molecules in which they are contained. When placed in a magnetic field, NMR active nuclei (such as 1H or 13C) absorb electromagnetic radiation at a frequency characteristic of the isotope. The resonant frequency, energy of the absorption and the intensity of the signal are proportional to the strength of the magnetic field. NMR spectroscopy is used by chemists and biochemists to investigate the properties of organic molecules, though it is applicable to any nucleus possessing spin.

Nucleic Acids: Nucleic acids are biological molecules essential for life, and include DNA (deoxyribonucleic acid) and RNA (ribonucleic acid). A nucleotide is composed of a nucleobase (nitrogenous base), a five-carbon sugar (either ribose or 2'-deoxyribose), and one phosphate group. The nucleobases are macromolecules like cytosine, guanine, adenine (DNA and RNA), thymine (DNA) and uracil (RNA), abbreviated as C, G, A, T, and U, respectively.

Oligonucleotide: An oligonucleotide is a short nucleic acid polymer, typically with fifty or fewer bases.

Organic Electronics: Organic electronics, plastic electronics or polymer electronics, is a branch of electronics dealing with conductive polymers, plastics, or small molecules. It is called 'organic' electronics because the polymers and small molecules are carbon-based. This contrasts with traditional electronics (or metal electronics), which relies on inorganic conductors such as copper or silicon.

Organic solar cell: An organic photovoltaic cell (OPVC) is a photovoltaic cell that uses organic electronics, a branch of electronics that deals with conductive organic polymers or small organic molecules for light absorption and charge transport.

Pathway (Metabolic pathway): In biochemistry, metabolic pathways are series of chemical reactions occurring within a cell. In each pathway, a principal chemical is modified by a series of chemical reactions. Enzymes catalyze these reactions, and often require dietary minerals, vitamins, and other cofactors in order to function properly.

Photobleaching: Photobleaching is the photochemical destruction of a fluorophore. In microscopy, photobleaching may complicate the observation of fluorescent molecules, since they will eventually be destroyed by the light exposure necessary to stimulate them into fluorescing.

Phytoremediation: Phytoremediation describes the treatment of environmental problems (bioremediation) through the use of plants that mitigate the environmental problem without the need to excavate the contaminant material and dispose of it elsewhere. Phytoremediation consists of mitigating pollutant concentrations in contaminated soils, water, or air, with plants able to contain, degrade, or eliminate metals, pesticides, solvents, explosives, crude oil and its derivatives, and various other contaminants from the media that contain them.

Potentiometric titrations: Any acid-base titration may be conducted potentiometrically. Two electrodes, after calibration [to relate potential in millivolts (mV) to a pH value] are immersed in a solution of the analyte. One is an indicator electrode, selective for H_3O^+ and the other a stable reference electrode. The potential difference, which after calibration is pH, is measured after the successive addition of known increments of acid or base titrant. When a potentiometric titration is being performed, interest is focused upon changes in the emf of an electrolytic cell as a titrant of known concentration is added to a solution of unknown. The method can be applied to all titrimetric reactions provided that the concentration of at least one of the substances involved can be followed by means of a suitable indicator electrode. The critical problem in a titration is to recognize the point at which the quantities of reacting species are present in equivalent amounts.

Ref: http://www.towson.edu/~topping/Experiment5Chem210.doc

Potentiometry: Potentiometry passively measures the potential of a solution between two electrodes, affecting the solution very little in the process. The potential is then related to the concentration of one or more analytes. (http://tera.chem.ut.ee/~koit/arstpr/pot_en.pdf)

Raman spectroscopy: Raman spectroscopy is a spectroscopic technique based on inelastic scattering of monochromatic light, usually from a laser source. Inelastic scattering means that the frequency of photons in monochromatic light changes upon interaction with a sample. Photons of the laser light are absorbed by the sample and then reemitted. Frequency of the reemitted photons is shifted up or down in comparison with original monochromatic frequency, which is called the Raman effect. This shift provides information about vibrational, rotational and other low frequency transitions in molecules. Raman spectroscopy can be used to study solid, liquid and gaseous samples.

(http://content.piacton.com/Uploads/Princeton/Documents/Library/UpdatedLibrary/Raman_Spectroscopy_Basics.pdf &
http://www.uta.edu/rfmems/Teralum_old/a1980021p.pdf)

Ribosomal proteins/rRNA/16S rRNA: A ribosomal protein is any of the proteins that, in conjunction with rRNA, make up the ribosomal subunits involved in the cellular process of translation. Ribosomal ribonucleic acid (rRNA) is the RNA component of the ribosome, the enzyme that is the site of protein synthesis in all living cells. Ribosomal RNA provides a mechanism for decoding mRNA into amino acids and interacts with tRNAs during translation by providing peptidyl transferase activity. The tRNAs bring the necessary amino acids corresponding to the appropriate mRNA codon. 16S ribosomal RNA (or 16S rRNA) is a component of the 30S subunit of prokaryotic ribosomes. It is 1,542 nucleotides in length. The genes coding for it are referred to as 16S rDNA and are used in reconstructing phylogenies.

SDS-PAGE electrophoresis: The separation of macromolecules in an electric field is called electrophoresis. A very common method for separating proteins by electrophoresis uses a discontinuous polyacrylamide gel as a support medium and sodium dodecyl sulfate (SDS) to denature the proteins. The method is called sodium dodecyl sulfate polyacrylamide gel electrophoresis (SDS-PAGE). The most commonly used system is also called the Laemmli method after U.K. Laemmli, who was the first to publish a paper employing SDS-PAGE in a scientific study. SDS-PAGE, sodium dodecyl sulfate polyacrylamide gel electrophoresis, is a technique widely used in biochemistry, forensics, genetics and molecular biology to separate proteins according to their electrophoretic mobility (a function of length of polypeptide chain or molecular weight).

Ref: http://www.ruf.rice.edu/~bioslabs/studies/sds-page/gellab2.html

Sensor array: A sensor array is a set of several sensors used to gather information. Usually the sensors and sensed phenomenon are both directional in nature, but this isn't always the case. Sometimes the data varies by position but not by direction.

Sorption processes: Sorption refers to the action of absorption or adsorption: Absorption, in chemistry, is a physical or chemical phenomenon or a process in which atoms, molecules, or ions enter some bulk phase - gas, liquid, or solid material. This is a different process from adsorption, since molecules undergoing absorption are taken up by the volume, not by the surface (as in the case for adsorption).

- Absorption is the incorporation of a substance in one state into another of a different state (e.g., liquids being absorbed by a solid or gases being absorbed by a liquid).
- Adsorption is the physical adherence or bonding of ions and molecules onto the surface of another phase (e.g., reagents adsorbed to solid catalyst surface).

Thermodynamics: Thermodynamics is a physical science that studies the effects on material bodies, and on radiation in regions of space, of transfer of heat and of work done on or by the bodies or radiation. It interrelates macroscopic variables, such as temperature, volume and pressure, which describe physical properties of material bodies and radiation, which in this science are called thermodynamic systems.

Therapeutic: Pertaining to the treating or curing of disease; curative. (http://dictionary.reference.com/browse/therapeutic).

Thick film technology (TFT): Thick film technology is used to produce electronic devices such as surface mount devices, hybrid integrated circuits and sensors. 'Thick film' (more correctly 'printed-and-fired') technology, uses conductive, resistive and insulating pastes containing glass frit, deposited in patterns defined by screen printing and fused at high temperature onto a ceramic substrate. The films are typically in the range 5–20μm thick, the range of resistivities is 10? /square to 10M? /square, there are considerable possibilities for building multi-layer structures.

Ultrasound spectrometry: Ultrasound spectrometry is based on detecting interactions of ultrasound wave with studied sample.

UV/VIS spectrometry: It adjacent (near-UV and near-infrared (NIR)) ranges. The absorption or reflectance in the visible range directly affects the perceived color of the chemicals involved. In this region of the electromagnetic spectrum, molecules undergo electronic transitions.

Viral nucleotides: Nucleotides present in virus molecules is called viral nucleotides.

Voltage sensitive dye: Voltage-sensitive dyes, also known as potentiometric dyes, ar e dyes which change their spectral properties in response to voltage changes. They are able to provide linear measurements of firing activity of single neurons, large neuronal populations or activity of myocytes. Many physiological processes are accompanied by changes in cell membrane potential which can be detected with voltage sensitive dyes. Measurements may indicate the site of action potential origin, and measurements of action potential velocity and direction may be obtained.

Voltammetry: Voltammetry applies a constant and/or varying potential at an electrode's surface and measures the resulting current with a three electrode system. This method can reveal the reduction potential of an analyte and its electrochemical reactivity. This method in practical terms is nondestructive since only a very small amount of the analyte is consumed at the two-dimensional surface of the working and auxiliary electrodes. In practice the analyte solutions is usually disposed of since it is difficult to separate the analyte from the bulk electrolyte and the experiment requires a small amount of analyte. The electrochemical cell, where the voltammetric experiment is carried out, consists of a working (indicator) electrode, a reference electrode, and usually a counter (auxiliary) electrode. In general, an electrode provides the interface across which a charge can be transferred or its effects felt. Because the working electrode is where the reaction or transfer of interest is taking place, whenever we refer to the electrode, we always mean the working electrode. The reduction or oxidation of a substance at the surface of a working electrode, at the appropriate applied potential, results in the mass transport of new material to the electrode surface and the generation of a current.

Ref: http://www.prenhall.com/settle/chapters/ch37.pdf

Voltammograms: A plot of the outcome of a voltammetry analysis.

Vortexing: A vortexing is a spinning, often turbulent, flow of fluid. Any spiral motion with closed streamlines is vortex flow. The motion of the fluid swirling rapidly around a center is called a vortex. Vortexing of biological substances or mixing of biological substances is d one in vortex mixer.

X–ray microtomography: Microtomography (commonly known as Industrial CT Scanning), like tomography, uses x-rays to create cross-sections of a 3D-object that later can be used to recreate a virtual model without destroying the original model. The term micro is used to indicate that the pixel sizes of the cross-sections are in the micrometer range. These pixel sizes have also resulted in the terminology micro-computed tomography, micro-ct, micro-computer tomography, high resolution x-ray tomography, and similar terminologies. All of these names generally represent the same class of instruments.

Ref: http://www.ami.ac.uk/courses/topics/0255_tft/index.html

J Biochem Tech (2010) 2(5):S1-S2
ISSN: 0974-2328

Utilization of high performance liquid chromatography for determination of antioxidant capacity

Jiri Sochor, Petr Babula, Boris Krska, Ales Horna, Ivo Provaznik, Jaromir Hubalek, Rene Kizek*

Received: 25 October 2010 / Received in revised form: 13 August 2011, Accepted: 25 August 2011, Published: 25 October 2011
© Sevas Educational Society 2011

Abstract

In this study, we are interested in possibility of determination and expression of biological value of twenty perspective genotypes of apricot (*Prunus armeniaca* L.). Work focuses on verification and validation of HPLC technique with electrochemical detector (ED) for determination of total antioxidant capacity (TAC), which may represent suitable method for comparison of biological value of fruits.

Keywords: Total antioxidant capacity, HPLC, *Prunus armeniaca* L.,

Jiri Sochor, Boris Krska

Department of Breeding and Propagation of Horticultural Plants, Faculty of Horticulture, Mendelu, Valticka 337, 691 44 Lednice, CZ

Petr Babula

Department of Natural Drugs, University of Veterinary and Pharmaceutical Sciences, Palackeho 1-3, CZ-612 42 Brno,

Ales Horna

Tomas Bata University, T.G. Ma saryka 275, 762 72 Zlin, CZ,

Ivo Provaznik

Department of Biomedical Engineering, VUT Brno, Kolejni 4, 61200 Brno, CZ,

Jaromir Hubalek

Department of Microelectronics, VUT Brno, Udolni 53, 602 00 Brno, CZ,

Rene Kizek[*]

Department of Chemistry and Biochemistry, Faculty of Agronomy, Mendelu, Zemedelská 1, 613 00 Brno, CZ

*Tel: +420 545 133 350, Fax: +420 545 212 044
 E-mail: kizek@sci.muni.cz

Introduction

Due to chemical variability of compounds with antioxidant capacity present in fruits, content of individual compounds is usually not known at all. Detection of biologically active compounds in biological matrix is usually very complicated, determined quantitative values are usually different and present compound/compounds responsible for antioxidant properties are not identified both using effective separation analytical methods and instruments. In addition, levels of individual antioxidants present in fruits need not necessarily correspond to total antioxidant capacity/activity (Adam et al. 2007). Main problem in this field is represented by absence of ideal analytical methods and techniques. Due to fact that antioxidants very easily participate in redox reactions, they are very suitable for electrochemical detection. Electrochemical determination of total antioxidant capacity is one of the possibilities how to express biological and nutrition value of fruits (Gazdik et al. 2008).

Materials and methods

Total antioxidant capacity was monitored using HPLC technique with coulometric detection, which represents one of the most sensitive detection techniques (Beklova et al. 2008). Concurrently, HPLC profile of fifteen flavonoids was detected.

HPLC-ED system consisted of two chromatographic pumps, Model 582 ESA (ESA Inc., Chelmsford, MA) (working range 0.001-9.999 ml min^{-1}) and reversed chromatographic column Zorbax SB C18 (150 × 4.6; size of particles 5 µm, Agilent Technologies, USA). For UV, UV detector Shimadzu (Model 528, ESA, USA) was used. For electrochemical detection, twelve-channel detector CoulArray (ESA, USA) with flow analytical cell (Model 6210, ESA, USA) was used. Flow cell contains planar glassy carbon electrode, hydrogen-palladium electrode as reference electrode and carbon electrode as auxiliary electrode. Sample was injected automatically by the use of autosampler (Model 542, ESA, USA), which has incorporated thermostat place for column. Column was thermostated to 30°C. Output data were processed by application CSW 32 software (Version 1,2,4, Data Apex, Czech Republic).

Real samples were prepared by graining of apricots in mortar with mixture of 50% methanol in water (v/v). After that homogenate was vortexed at 20 minutes under ambient temperature. Supernatant was obtained by centrifugation under 14.000 rpm, 10 minutes and then directly injected to HPLC.

Chromatographic conditions were optimized: volume of injection of standard mixtures and real samples 30 µl, mobile phase A consisted of formic acid (0.2 %, v/v), mobile phase B was acetonitrile. Profile of gradient was linearly increased from 12 to 22 % for B (v/v) (0 - 20 min), to 50 % B (20 – 25 min), to 55 % (25 – 30 min). Flow rate was 0.8 µl.min^{-1}. Electrochemical detector scanned responses at potentials -80, 0, 80, 160, 240, 320, 400, 480, 560, 640, 720 and 800 mV. Resulting detection was expressed in microcoulombs. Mathematical and statistical analysis of experimental data was carried out in package MATLAB®, Version 7.9.0.529 (R2009b).

Results and discussion

For determination of total antioxidant capacity, HPLC-ED technique with gradient elution was used for determination of flavonoids in plant samples. Based on differences of signal areas, comparative series of determined fruits, where differences in contents of antioxidant flavonoids are expressed as relative % and interpreted as relative antioxidant capacity, were investigated.

Lignite is able to absorb high amount of water, in mined state contains 50% at least. This ability is reversible during processes of drying and hydratation. Natural mineral zeolite is additive soil substance of volcanic origin (tetragonal sodium aluminosilicate), which contains approximately 70 % of silicon oxide. Zeolite is highly porous with ability of water and ions absorption with subsequent releasing based on ions exchange (Diopan et al. 2008).

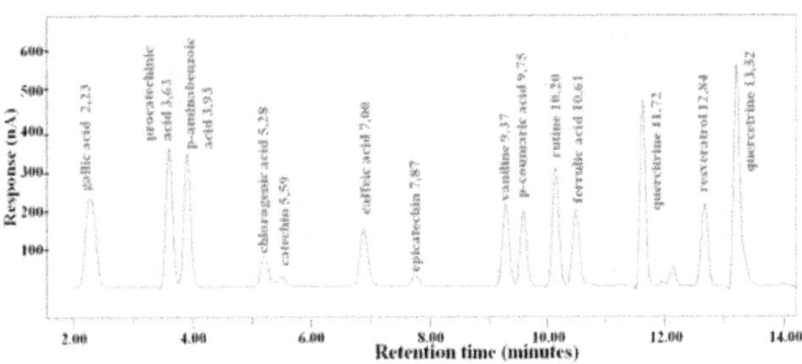

Figure 1: Chromatogram of cultivars LE-9299. HPLC profile of fifteen flavonoids

We also determined total antioxidant capacity of fruits of individual apricot cultivars related to content of major represented antioxidant compounds; quantification of present antioxidants with their participation on TAC was also possible.

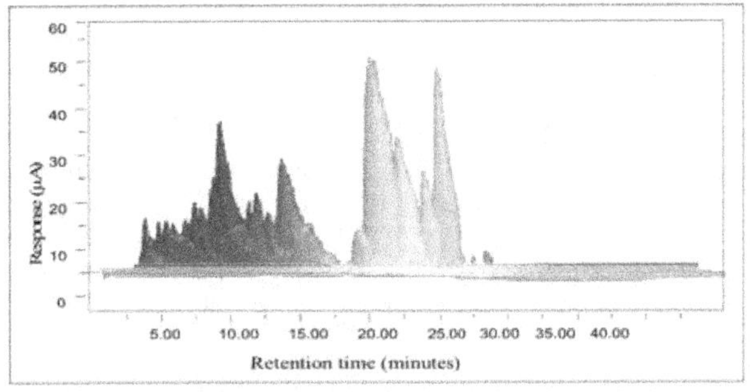

Figure 2: Output of CoulArray detector with twelve working electrodes in the case of cultivars LE-9299Conclusion

Our results (Fig 1 & Fig 2) demonstrate variation of antioxidant capacity in dependence on type of phenolic compounds present in fruits as well as variation between individual types of phenolic compounds – some types of phenolics demonstrate higher antioxidant in comparison with others (Diopan et al. 2008). It is supposed that on protective effect participates possibility of plant polyphenolics to scavenge reactive oxygen radicals, which are able to generate highly reactive hydroxyl radicals (Adam et al. 2007; Beklova et al. 2008). Due to direct connection between antioxidant activity/capacity and ability of compound to be oxidized/reduced, it means due to ability to provide signal in electrochemical detection (Gillman et al. 1995), connection of selective and very sensitive electrochemical detection with high performance liquid chromatography (HPLC-ED) enabling simultaneous separation of many constituents in one run represents ideal analytical tool for solving of these very difficult questions.

Conclusion

We developed suitable separation procedure and technique for determination of total antioxidant capacity in newly cultivated apricot cultivars. HPLC/ED enabled monitoring of reaction kinetics at applied potential, evaluation of structures of antioxidant complexes and their total antioxidant capacity.

Acknowledgement

This work was financially supported by grants NAZV 91A032 a GAČR 102/08/1546.

References

Adam V, Mikelova R, Hubalek J, Hanustiak P, Beklova M, et al. (2007) Utilizing of square wave voltammetry to detect flavonoids in the presence of human urine. Sensors 7(10):2402-2418

Beklova M, Zitka O, Gazdik Z, Adam V, Hodek P, et al. (2008) Electroanalytical techniques for determination of flavonoids. Toxicology Letters 180:S230-S230

Diopan V, Shestivska V, Adam V, Macek T, Mackova M, et al. (2008) Determination of content of metallothionein and low molecular mass stress peptides in transgenic tobacco plants. Plant Cell Tissue and Organ Culture 94(3):291-298

Gazdik Z, Krska B, Adam V, Sa loun J, Pokorna T, et al. (2008) Electrochemical Determination of the Antioxidant Potential of Some Less Common Fruit Species. Sensors 8(12):7564-7570

Gillman MW, Cupples LA, Gagnon D, Posner BM, Ellison RC, et al. (1995) Protective effect of fruits and vegetables on development of stroke in men. Jama-Journal of the American Medical Association 273(14):1113-1117

J Biochem Tech (2010) 2(5):S3-S4
ISSN: 0974-2328

Retention behavior of oligonucleotides isomers studied by means of hydrophilic interaction chromatography

Hana Fuksova, Blanka Hegrova, Pavel Kostecka, Libuse Trnkova*, Miroslava Bittova

Received: 25 October 2010 / Received in revised form: 13 August 2011, Accepted: 25 August 2011, Published: 25 October 2011
© Sevas Educational Society 2011

Abstract

Short oligonucleotides and its isomers are deeply studied for better understanding of nucleic acid properties and interactions. High Performance Liquid Chromatography (HPLC) provides suitable tool for their characterization. Our interest in this study was focused on the separation of pentamers, consisted of adenine, guanine and thymine by means of hydrophilic liquid chromatography mode (HILIC). The influence of experimental parameters such as mobile phase pH and/or buffer concentration, percentage of organic solvent in mobile phase and flow rate was evaluated and the most acceptable conditions were selected for further analysis. Finally, obtained results were compared with commonly applied Reverse Phase High Performance Liquid Chromatography (RP–HPLC).

Keywords: Hydrophilic interaction chromatography, oligonucleotides

Introduction

Liquid chromatography (LC) belongs to the most applied separation analytical methods in bioanalysis. Based on its general principles, HILIC mode affords alternative information to RP–HPLC separation (Gilar et al. 2002; Wang et al. 2005). The main goal of this work was the evaluation of relationship between oligonucleotide sequence and retention time in HILIC under various experimental conditions, consequential selection of the most suitable conditions for HILIC separation and comparison of results with RP– HPLC analysis.

Hana Fuksova Blanka Hegrová, Libuse Trnkova*, Miroslava Bittova

Masaryk University, Faculty of Science, Department of Chemistry, Kotlarska 2, 611 37 Brno, Czech Republic

Pavel Kostecka

Institute of Biophysics ASCR, v.v.i., Kralovopolska 135, 612 65 Brno, Czech Republic

*Tel: +420 549 497 754, Fax: +420 541 211 214
E-mail: libuse@chemi.muni.cz

Materials and methods

All experiments were carried out on Shimadzu 10 AVP HPLC system with diode–array detector and ZIC–pHILIC (100 x 4.6mm, 5 μm) and Kinetex C18 (100 x 3 mm, 2.6 μm) chromatographic columns were used. Acetonitrile/ammonium carbonate mobile phase in HILIC mode and methanol/ammonium acetate mobile phase in RP mode were tested.

Results and Discussion

Equal base composition but unequal polynucleotide sequence results in various affinity to chromatographic column stationary phases and thus the analysis in different chromatographic systems leads to comprehensive results acquirement. Retention behavior of selected pentamers was studied under different conditions by HILIC and it was confirmed that analytes separated by use of HILIC elute in different order in comparison to RP–HPLC (Fig 1).

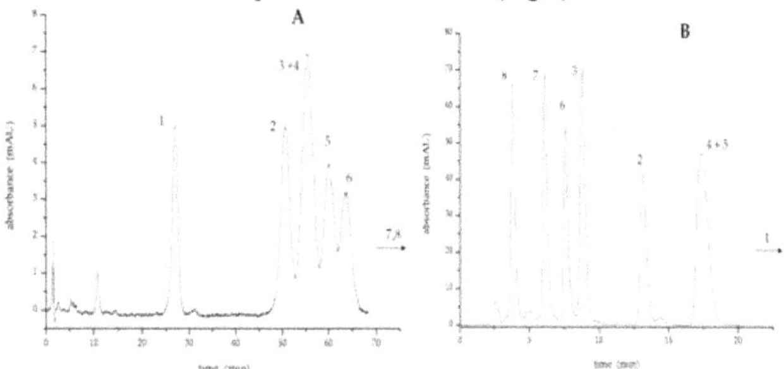

Figure 1: Separation of pentamer mixture using **A**, HILIC and **B**, RP–HPLC. 1 – ATTTT, 2 – ATGTT, 3 – ATGAT, 4 – AGTTT, 5 – AGTTA, 6 – AGATT, 7 – AGTAG, 8 – AGAGT

Conclusion

It was shown that both mentioned LC modes can be applied in short oligonucleotides separation. The most suitable conditions for HILIC separation were compared with RP–HPLC. However, RP–HPLC gives more sensitive results in shorter times and thus is more suitable to be used in this field of study.

Acknowledgement

This work was supported by the Grant Agency of the Czech Republic, project nr. 203/08/P598 and by the Ministry of Education, Youth and Sports of the Czech Republic, project LC06035.

References

Gilar M, Fountain KJ, Budman Y, Neue UD, Yardley KR, et al. (2002) Ion-pair reversed-phase high-performance liquid chromatography analysis of oligonucleotides: Retention prediction. Journal of Chromatography A 958(1-2):167-182

Wang XD, Li WY, Rasmussen HT (2005) Orthogonal method development using hydrophilic interaction chromatography and reversed-phase high-performance liquid chromatography for the determination of pharmaceuticals and impurities. Journal of Chromatography A 1083(1-2):58-62

J Biochem Tech (2010) 2(5):S5-S6
ISSN: 0974-2328

CITP analysis of aminopolycarboxylic acids

Jan Brezina, Radek Sevcik, Premysl Lubal*

Received: 25 October 2010 / Received in revised form: 13 August 2011, Accepted: 25 August 2011, Published: 25 October 2011
© Sevas Educational Society 2011

Abstract

This contribution concerns the analytical procedure of industrial samples of aminopolycarboxylic acids by means of capillary izotachophoresis (CITP).

Keywords: electromigration techniques, separation, aminopolycarboxylic acids

Introduction

CITP is an analytical electromigration technique which enables ion's separation based on their different mobilities and it can be used for determination of ionic analytes. A polyamino carboxylic acid (complexone) is a compound containing one or more nitrogen atoms connected through carbon atoms to one or more carboxyl groups. Polyamino carboxylates, which have lost acidic protons, form strong complexes with metal ions by donation of electron pairs from the nitrogen and oxygen atoms to the metal ion to form multiple chelate rings. This property makes polyamino carboxylic acids useful in a wide variety of chemical, medical and environmental applications (Anderegg et al. 2005). The goal of this work was to verify the possible application of this technique for the migration study of zwitterionic compounds and their determination in a mixture and in industrial samples.

Materials and Methods

Method optimization for separation and determination of analytes were carried out on electrophoretic equipment EA 102 (Villa Labeco, Spišská Nová Ves, Slovakia) on PTFE capillary (diameter 0.3 mm, length 90 mm) joined with conductivity detector under

Jan Březina, Radek Ševčík, Přemysl Lubal*

Department of Chemistry, Faculty of Science, Masaryk University, Brno, Czech Republic.

*Tel: +420 549 49 5637, Fax: +420 541 211 214
 E-mail: lubal@chemi.muni.cz

laboratory temperature. Compounds (injection 30 µl) were analyzed in the following system: 20 mM HCl + glycylglycine (leading electrolyte, pH = 3.1), 20 mM acetic acid + NH_3 (terminating electrolyte, pH = 4.80). 0.5% hydroxyethylcellulose solution was added in order to eliminate electroosmotic flow.

Results and Discussion

The procedure was optimized by analysis of aminopolycarboxylates of known concentration in the mixture (Fig 1 & Fig 2). The optimal pH = 3.1 was determined by mobility measurement in region pH = 2.5–4.5 since no formation of mixed zones was observed. Detection limits of chosen analytes were determined from the calibration curves in 0.1–0.8 mM concentration region: NTA (0.058 mM), EDTA (0.052 mM), IDA (0.082 mM), glycolic acid - GlyA (0.097 mM). This new analytical procedure was employed in the analysis of real industrial samples (Syntron A, Syntron B – Draslovka, Trilon A, Trilon B – BASF, Ferrazone – Akzo Nobel).

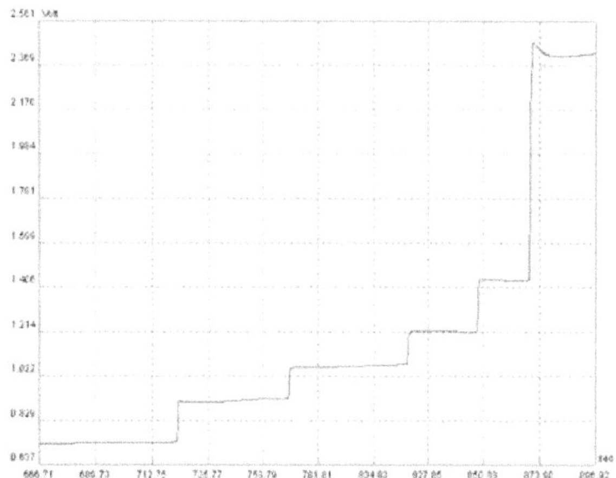

Figure 1: The ITP analysis of mixture of aminopolycarboxylates (c = 0.5 mM, left): NTA, EDTA, IDA, GlyA – the order of analytes in CITP record).

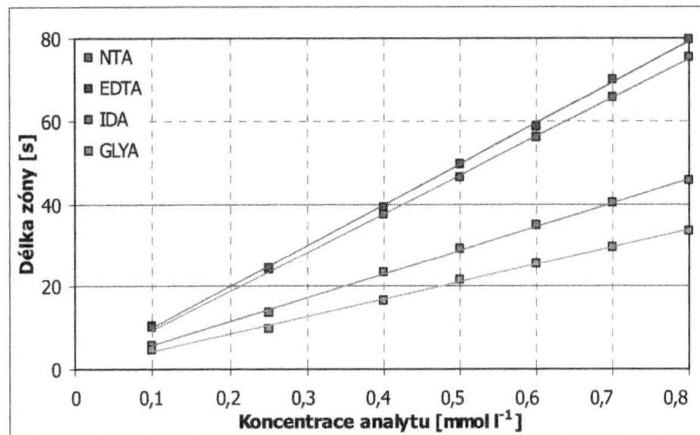

Figure 2: Calibration plots for analyzed samples

The mobilities of species of studied analyses were calculated from pH dependences and their estimates were used for simulation of ITP record (Fig 3). Comparing with experimental record (Fig 1 & Fig 2), the agreement is evident.

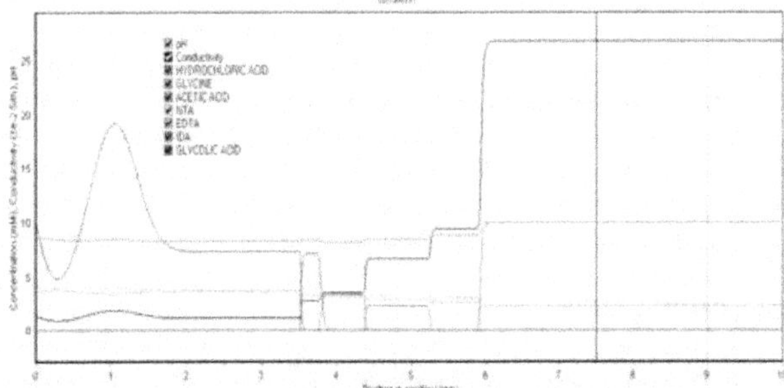

Figure 3: The CITP record of mixture of analytes (c = 4 mM) simulated by means of SIMUL v 5.0 software (Bohuslav Gaš et al. 2007).

Conclusion

This contribution is focused on optimization and determination of aminopolycarboxylic acids in industrial samples by means of CITP which is suitable alternative to other analytical methods.

Acknowledgement

The work was supported by Ministry of Education of the Czech Republic (projects KONTAKT ME09065 and LC06035).

References

Anderegg G, Arnaud-Neu F, Delgado R et al. (2005) Critical evaluation of stability constants of metal complexes of complexones for biomedical and environmental applications. Pure Appl Chem 77(8):1445–1495

Bohuslav Gaš et al. (2007) SIMUL v 5.0. Available via DIALOG. http://web.natur.cuni.cz/gas/. Cited 15 Oct 2010

J Biochem Tech (2010) 2(5):S7-S8
ISSN: 0974-2328

Determination of sarcosine as possible tumour marker of prostate tumours

Natalia Cernei, Michal Masarik, Jaromir Gumulec, Ondrej Zitka, Petr Babula, Rene Kizek

Received: 25 October 2010 / Received in revised form: 13 August 2011, Accepted: 25 August 2011, Published: 25 October 2011
© Sevas Educational Society 2011

Abstract

Amino acid sarcosine, known also as N-methylglycine, may be established as new very important marker in prostate malignant tumours and may be determined by very simple test. Cancer of prostate is one of the most incident types of malignant tumours in men. More than one thousand men in Czech Republic die due to this disease. As well as in the case of other malignant tumours, for initiation of treatment well timed diagnosis of disease is necessary. For determination of sarcosine we employed the ionex chromatography with postcolumn derivatization by ninhydrin. We achieved the calibration curve linearity R^2=0.9984 with limit of detection 500nM. Moreover we confirmed that the calibration was not affected by presence of another aminoacids and ensure that selectivity of separation is near to 99% efficiency.

Keywords: sarcosine, new marker of prostate cancer, cysteine and homocysteine

Introduction

Sarcosine is naturally occurring non-toxic amino acid soluble in water. Sarcosine may be placed into group of biogenic amines, which occurs in different types of foods (fish, meat, beer, wine, cabbage, olives). In foods, sarcosine is originated during process of proteins fermentation by action of various kinds of organisms. At its consummation, sarcosine (Fig 1) passes to urine in unchanged form. Sarcosine also originates in liver and kidneys as intermediate of choline metabolism. At the beginning of 2009, study demonstrating diagnostic potential of this amino acid was published in very prestigious journal Nature. Sarcosine is very reputable marker also in early stages of tumours. Very important is also fact that sarcosine was not present in urine of healthy people. Due to this fact, probability of false positivity of investigation is minimized. In accordance with published results, sarcosine is significantly preferable marker of prostate cancer to prostate-specific antigene, whose presence in blood is routinely analyzed. Presence of amino acid sarcosine indicates rightly aggressive, high-malignant tumours. Aim of our work consisted in designation of chromatographic method for determination of sarcosine in presence of free amino acids as cysteine and homocysteine.

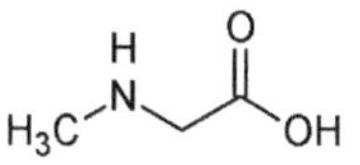

Figure 1: Structure of sarcosine

Natalia Cernei, Ondrej Zitka, , Rene Kizek*

Department of Chemistry and Biochemistry Mendel University, Zemedelska 1, CZ-613 00 Brno, Czech Republic

Michal Masarik, Jaromir Gumulec

Department of Pathological Physiology, Faculty of Medicine, Masaryk University, Komenskeho namesti 2, CZ-662 43 Brno, Czech Republic

Petr Babula

Department of Natural Drugs, Faculty of Pharmacy, University of Veterinary and Pharmaceutical Sciences, Palackeho 1-3, CZ-612 42 Brno, Czech Republic

*Tel: +420 545 133 350, Fax: +420 545 212 044
E-mail: kizek@sci.muni.cz

Materials and methods

Sarcosine of purity 99 % was obtained from Fluka BioChemika (Switzerland). Solution of sarcosine for preparing of calibration curve was prepared in buffer Na:TDG (N_3Na - 0.10 g, NaCl-11.5 g, $C_6H_8O_7$-14 g in 1l of H_2O). Separation of sarcosine was carried out by the help of ionex chromatography on apparatus Amino Acid Analyzer AAA 400 (Ingos, Czech Republic). AAA analyzer works on basis of medium pressure liquid chromatography with ionex column and ninhydrin postcolumn derivatization. Glass column with inner diameter 3.7 mm and 350 mm length was filed manually with strong catex ionex in natrium cycle LG ANB with approximately 12μm particles and porosity 8%. Column was tempered in range 35 - 95°C. Double channel VIS detector with inner cell volume 5 μl worked statically under two wavelengths 440 and 570nm. Solution of ninhydrin (Ingos, Czech Republic) was prepared by dilution in 75 % v/v metylcelosolve (Ingos, Czech Republic) and in 25 % v/v 4M acetic buffer (pH5.5). For reduction $SnCl_2$ (Lachema, Czech Republic) was used. Prepared solution of ninhydrin was stored under inert atmosphere (N_2) in dark and cooled. Elution of Sarcosine was done by buffer contains Citric acid 11.11g, natrium citrate 4.04g and NaCl 9.25g x1L^{-1}. Flow rate was 0.3 ml/min. Reactor temperature was 120°C. Conductivity of used miliQ water was 18M? (Jamaspishvili et al. 2010).

Results and discussion

In experimental part, firstly we were focused on analytical determination of sarcosine. Ionex chromatography was chosen as the most suitable analytical technique. Sarcosine was analyzed in tested concentration range from 5 to 1000 μM with error of determination of about 8%. In chromatograms very well developed and separated chromatographic signals at retention time of 32 min were obtained. Calibration curve was linear in entire tested range with equation $y=0.0266x-0.2794$ $R^2=0.9984$ (Fig 2). Limit of detection of sarcosine was determined as 3S/N and was about 500 nM. Limit of quantification of sarcosine was determined as 10 S/N - 5 μM.

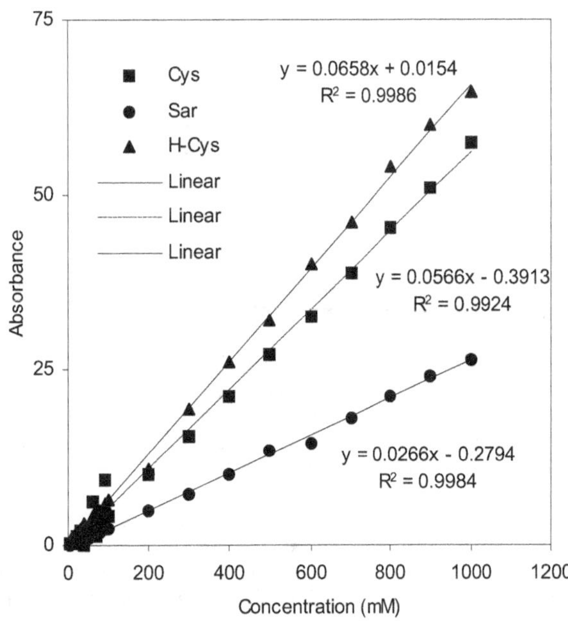

Figure 2: Calibration curve of sarcosine in presence of cysteine and homocysteine

Analysis of sarcosine in presence of two very hardly determinable amino acids (cysteine and homocysteine) was carried out in subsequent experiments. Presence of these two amino acids had no effect on chromatographic signals of sarcosine. All chromatographic signals were well developed and separated. Moreower free amino acids are present in biological samples – due to this fact it is absolutely necessary to propose suitable technique for determination of sarcosine in presence of free amino acids. In our next experiment, sarcosine at concentration 10 μM was added into mixture of 17 amino acids of concentration 20 μM. In Fig 3, typical chromatographic record is demonstrated. Individual amino acids are separated very well; no interferences with determined sarcosine were observable.

Conclusion

Chromatographic method enabling simultaneous determination of sarcosine in presence of 17 free amino acids was designed in our experimental work.

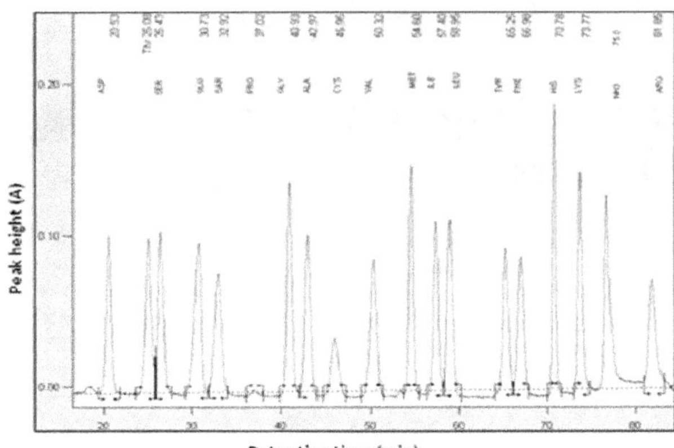

Figure 3: Typical chromatographic record of common amino acids with addition of sarcosine

Acknowledgements

The work has been supported by grants: GACR 301/09/P436, IGA MZ 10200-3 and NANOSEMED GA AV KAN208130801

References

Jamaspishvili T, Kral M, Khomeriki I et al. (2010) Urine markers in monitoring for prostate cancer. Prostate Cancer Prostatic Dis. 13(1):12-9

J Biochem Tech (2010) 2(5):S9-S10
ISSN: 0974-2328

Determination of GSH and GSSG in blood plasma of patients by HPLC-ED

Ondrej Zitka, Natalia Cernei, Vojtech Adam, Ales Horna, Rene Kizek*

Received: 25 October 2010 / Received in revised form: 13 August 2011, Accepted: 25 August 2011, Published: 25 October 2011
© Sevas Educational Society 2011

Abstract

Thanks to many chemicals and physical stresses generated by outside environment there still is unpredictable danger of damaging of organism by reactive oxygen species (ROS). Biological harmfulness of ROS is generally given by the subsequent oxidation of essential cellular structures. Organism is defending against ROS by synthesis of small molecules which can neutralize radicals. The most suitable existing organic molecule widespread in whole organism which can decrease effect of ROS is glutathione (GSH). Therefore we interested in changes of GSH levels by oncological patients. We optimized fast and robust method for analysis blood plasma of patients by employment of HPLC with coulometric detection. Our results based on hunderts of analyzed samples shows significant oxidation damage by disbalance between GSH/GSSG ratio.

Keywords: reactive oxygen species, glutathione, HPLC with electrochemical detection.

Introduction

If oxidative stress in form of ROS is occuring in the organism mechanisms which can stabilize homeostasis are employed. One of well described mechanism is stabilisation of homeostasis by synthesis of tripeptide glutathione (GSH) which consists of aminoacids glutamine, cysteine and glycine (Obr. 1). It can be find in tissues and body liquids of animals and also in plants. Neutralization of ROS by GSH is possible by cascade of reactions which are called as Glutathione-acorbate cycle (Noctor and Foyer 1998). On the end of cycle oxidized form of glutathione (GSSG) and reduced form of previous danger radical are produced. Therefore observing of concentration levels between both forms of glutathione GSH and GSSG might be very useful for examination of state of oxidation damaging of whole organism. Concentration of GSH and GSSG in real samples as blood or tobacco cells extract have to be analyzed by various methods as Capilary electrophoresis with fluorescenc detection after selective derivatization (Maeso et al. 2005; Zhang et al. 2005). This kind of approach is very sensitive especially if laser induced florescence (LIF) detector is used. Moreover if direct detection after separation step must be preferred electrochemical detector is the most sensitive an selective (Hiraku et al., 2002). Both methods of detection LIF for electrophoresis or electrochemical detection by HPLC has unlimited benefits in field of miniaturization in the near future. Anyway gluthation is are reactive compound which free level in biological matrix could be strongly affected by bonding to other proteins (Fratelli et al. 2002) and thus this problematic is firther disputed (Rossi et al. 2002).

Materials and Methods

Aim of our work was to optimize method which might be useful for determination of GSH and GSSG. These compounds of interests are involved in gluthation-ascorbate cycle and thus we asume that could be good representing in determination of oxidative damage of organism.

We analyzed more than 7 hundreds patients blood plasma. Samples of blood serum of cancer patients were denatured in 5% trifluoroacetic acid and after centryfugation supernatant were injected in chromatographic system. For purpose of separation and detection of GSH and GSSG gradient chromatographic system was employed. System consisted of two chromatographic pumps, autosampler and multichannel coulometric electrochemical detector Coularray.

HPLC-ED system consists of two chromatographic pumps Model 582 ESA (ESA Inc., Chelmsford, MA) (working range 0.001-9.999 ml min^{-1}) and chromatographic column with reverse phase Zorbax eclipse AAA C18 (150 × 4.6; 3,5 μm particles, Agilent

Ondrej Zitka, Natalia Cernei, Vojtech Adam, Rene Kizek*

Department of Chemistry and Biochemistry, Faculty of Agronomy, Mendel University in Brno, Zemedelska 1, CZ-613 00 Brno, Czech Republic

Ales Horna

Radanal Ltd., Okruzni 613, CZ-530 03, Pardubice, Czech Republic

*Tel: +420 545 133 350, Fax: +420 545 212 044
E-mail: kizek@sci.muni.cz

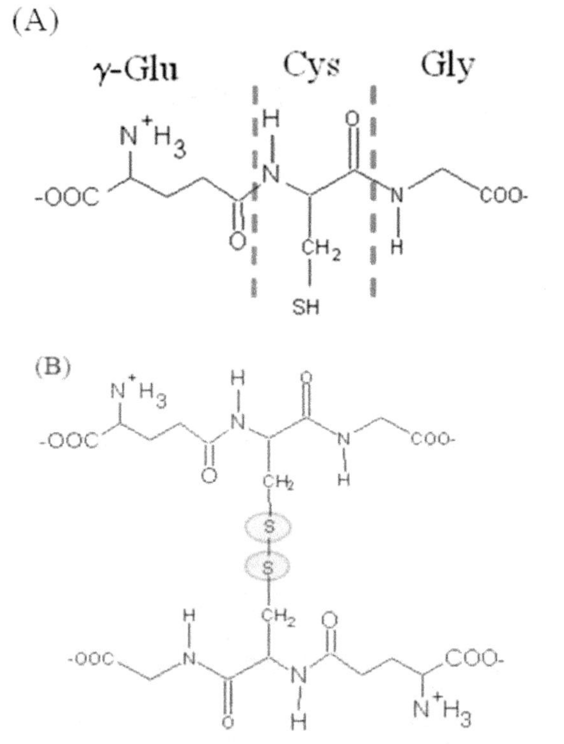

(A)

(B)

Figure 1: Structure of glutathione reduced form (A) and gluthatione – oxidized form (B).

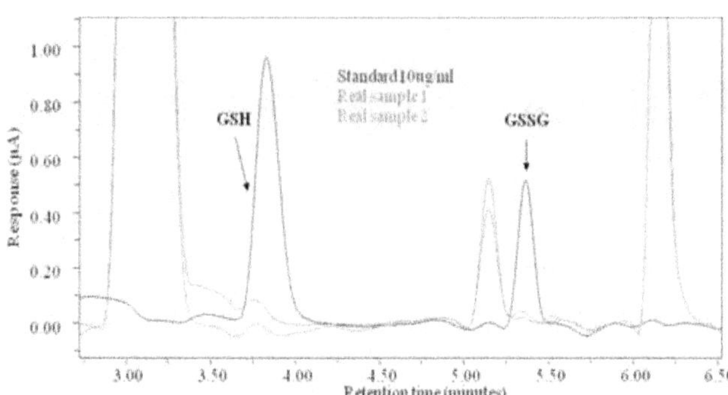

Figure 2: Chromatograms of standard of GSH and GSSG in overlay with two real samples obtained from blood plasma of two patients

Technologies, USA) and twelve-channel CoulArray electrochemical detector (Model 5600A, ESA, USA). Detector consists of three flow analytical chambers (Model 6210, ESA, USA). Each chamber contains four analytical cells. One analytical cell contains two referent (hydrogen-paladium), and two counter and one porous graphite working electrodes. Electrochemical detector is situated in control module which is thermostated. Sample (20 μl) was injected by autosampler (Model 542, ESA, USA), which has thermostated space for column. Column was termostated at 30°C. Flow rate of mobile phase was 1 ml min^{-1}. Mobile phase consists of A: trifluoroacetic acid (80 mM) a B: 100% Met-OH. Compounds were eluted by following linear increasing gradient: 0-2 min (3%B), 2->4 min (10%B), 6->10 min (40%B), 10-12 min (98%B). Detection was carried out at applied potential 900mV.

Results and discussion

There was some scientific interest with regards to exclusively option of bond between GSH and cis-Pt in treated organism (Ishikawa and Aliosman, 1993). Later these types of interaction were studied on theoretical computing (Zimmermann et al. 2005), and so by real in vitro experiment (Zitka et al. 2007). But for better understanding of behaving of glutathiones in live cells we tried to study only contents of them in cancer patients. We developed fast and rapid method for separation and electrochemical detection of GSH and GSSG. Time of one analysis was 20 minutes. Required sensitivity of detection of glutathiones is due to fact that thiol group is strongly electroactive at applied potential 900mV.

Conclusion

Our developed method was useful for analysis more than 7 hundreds

of patients. We observed that there was significant different level between concentrations of GSH and GSSG.

Acknowledgements

The work has been supported by GA ČR 522/09/0239, MSMT 6215712402 and 1M06030.

References

Fratelli M, Demol H, Puype M et al (2002) Identification by redox proteomics of glutathionylated proteins in oxidatively stressed human T lymphocytes. Proceedings of the National Academy of Sciences of the United States of America 99:3505-3510

Hiraku Y, Murata M, Kawanishi S (2002) Determination of intracellular glutathione and thiols by high performance liquid chromatography with a gold electrode at the femtomole level: comparison with a spectroscopic assay. Biochimica Et Biophysica Acta-General Subjects 1570:47-52

Ishikawa T, Aliosman F (1993) Glutathione-associated cis-diamminedichloroplatinum(ii) metabolism and atp-dependent efflux from leukemia-cells - molecular characterization of glutathione-platinum complex and its biological significance. Journal of Biological Chemistry 268:20116-20125

Maeso N, Garcia-Martinez D, Ruperez FJ et al. (2005) Capillary electrophoresis of glutathione to monitor oxidative stress and response to antioxidant treatments in an animal model. Journal of Chromatography B-Analytical Technologies in the Biomedical and Life Sciences 822:61-69

Noctor G, Foyer CH (1998) Ascorbate and glutathione: Keeping active oxygen under control. Annual Review of Plant Physiology and Plant Molecular Biology 49:249-279

Rossi R, Milzani A, Dalle-Donne I et al (2002) Blood glutathione disulfide: In vivo factor or in vitro artifact? Clinical Chemistry 48:742-753

Zhang JY, Hu ZD. Chen XG (2005) Quantification of glutathione and glutathione disulfide in human plasma and tobacco leaves by capillary electrophoresis with laser-induced fluorescence detection. Talanta 65:986-990

Zimmermann T, Zeizinger M, Burda JV (2005) Cisplatin interaction with cysteine and methionine, a theoretical DFT study. Journal of Inorganic Biochemistry 99:2184-2196

Zitka O, Huska D, Krizkova S et al (2007) An investigation of glutathione-platinum(II) interactions by means of the flow injection analysis using glassy carbon electrode. Sensors 7:1256-1270

J Biochem Tech (2010) 2(5):S11-S12
ISSN: 0974-2328

HPLC with electrochemical detection for studying of synthesis of phytochelatins in cadmium treated flax

Ondrej Zitka, Jitka Najmanova, Natalia Cernei, Vojtech Adam, Ales Horna, Rene Kizek*

Received: 25 October 2010 / Received in revised form: 13 August 2011, Accepted: 25 August 2011, Published: 25 October 2011
© Sevas Educational Society 2011

Abstract

If there would be free heavy metals occurring in the organism without control it initiates creation of number of dangerous species (ROS). These species are disturbing of homeostasis of organism by subsequent oxidation of essential cellular structures as nucleic acids. Animal and plant organism has both one efficient defending system which includes synthesis of glutathione (GSH). But plants are limited in case of escape from metal contaminated area. Therefore plant cell has mechanism includes synthesis of phytochelatins (PC) from GSH by phytochelatin synthase. PC´s are able to immobilize heavy metal ion via number of thiol groups. We treated the plants of flax (*Linum usativum*) by various concentration of Cd(II) ions (50, 250, and 500µM) in our study. Subsequently we used our developed method for determination of oxidation stress as GSH/GSSG ratio and levels of PC-2,3,4 and 5. We observed that most synthesised PC under the highest concentration of Cd(II) treatment was PC-2,4 and 5.

Keywords: Heavy metals, phytochelatins, electrochemical detection, reactive oxygen species.

Introduction

The world water and soil contamination is widespread problem which has limited scientific attention because it critically concerns the less developed countries mostly. Nowadays attention is paid to contamination with subsequent link to food chain – extensive pollution of the environment (Gawel et al. 1996; Gawel et al. 2001). The heavy metals are one of the most toxic groups of these undesirable compounds that threaten both plants and animals (Cobbett 2000b; Hall 2002; Zenk 1996).

The management of heavy metal ions inside the plant cell could by partially secured by thiols as glutathione and phytochelatins (Clemens 2001; Cobbett and Goldsbrough 2002; Cobbett 2000a).Phytochelatins ((γ-Glu-Cys)n-Gly) are molecules synthetized by PC-synthase from GSH which structure is (γ-Glu-Cys)-Gly. Name of PC is determined by number of repetitions (γ-Glu-Cys)n in PC structure (n...2,3,4,5 etc.). Then PC2, PC3, PC4 and PC5 could be created. The synthesis of phytochelatins is catalyzed by γ-Glu-Cys dipeptidyl transpeptidase named as phytochelatin synthase (PCS) which is crucial enzyme makes the plant resistant against heavy metal stress (Cobbett 1999; Vatamaniuk et al. 2001). High pressure liquid chromatography with (HPLC) with electrochemical detection (ED) is suitable method for studying of thiol substances due to presence sulfhydryl group SH which is very electro active.

Ondrej Zitka, Natalia Cernei, Vojtech Adam, Rene Kizek*

Department of Chemistry and Biochemistry, Faculty of Agronomy, Mendel University in Brno, Zemedelska 1, CZ-613 00 Brno, Czech Republic.

Jitka Najmanova

Departoment of Biochemistry and Microbiology, Institute of Chemical Technology Prague, Technicka 5, CZ-166 28, Prague, Czech Republic.

Ales Horna

Radanal Ltd., Okruzni 613, CZ-530 03, Pardubice, Czech Republic.

*Tel: +420 545 133 350, Fax: +420 545 212 044
E-mail: kizek@sci.muni.cz

Materials

For purpose of separation and detection of PC 2,3,4 and 5 gradient chromatographic system was employed. System consisted of two chromatographic pumps, autosampler and multichannel coulometric electrochemical detector Coularray. Samples ware prepared according our developed protocol. Single plants of flax which were treated by various concentration of Cd (50, 250, and 500µM) were homogenized by ultrasonic and resolved in 0.2 M phosphate buffer (pH 7.2) and after centrifugation supernatant were injected in chromatographic system.

Method

HPLC-ED system consists of two chromatographic pumps Model 582 ESA (ESA Inc., Chelmsford, MA) (working range 0.001-9.999 ml min^{-1}) and chromatographic column with reverse phase Zorbax eclipse AAA C18 (150 × 4.6; 3,5 µm particles, Agilent Technologies, USA) and twelve-channel CoulArray electrochemical

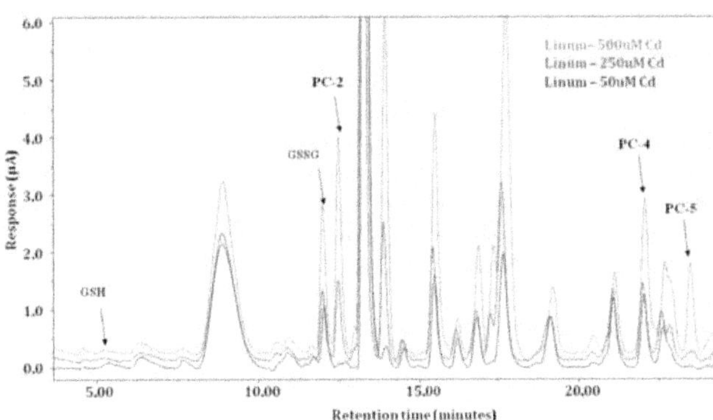

Figure 1: Structure of phytochelatin. Repetition of γ-glutamyl-cysteinyl could be n-2,3,4,5 and everytime is terminated by glycine.

detector (Model 5600A, ESA, USA). Detector consists of three flow analytical chambers (Model 6210, ESA, USA). Each chamber contains four analytical cells. One analytical cell contains two referent (hydrogen-palladium), and two counter and one porous graphite working electrode. Electrochemical detector is situated in control module which is thermostated. Sample (15 µl) was injected by autosampler (Model 542, ESA, USA), which has thermostated space for column. Column was termostated at 30°C. Flow rate of mobile phase was 1 ml min^{-1}. Mobile phase consists of A: trifluoric acid (80 mM) a B: 100% Met-OH. Compounds were eluted by following gradient: 0-7 min (3% B), 7->8 min (15 %B), 8-15 min (15 % B), 15->25 min (30 % B), 25-28 min (98 % B), 28-33 min (98 % B). Detection was carried out at applied potential 900mV.

Results and discussion

We developed fast and rapid method for separation and electrochemical detection of PC 2,4,5 and we were also able to detect amount of GSH and GSSG to. Time of one analysis was 45 minutes including regeneration of the column. Influence of treating by cadmium on generation of ROS is well observed on ratio of GSH and GSSG which is inversed than in normal state.

Figure 2: Chromatograms real samples of flax treated by 50, 250 and 500 µM cadmium.

Similarly the PC-2,4 and 5 are increasing with amount of cadmium. Interesting trend was observed in increase of PC-5 which is not commonly preferred form of PC in organism. Forming of PC-5 was increasing markedly due to 500mM cadmium. On the other way we can say that PC-4 could be in the plant even if not treated or if treated less than 50µM cadmium. But occurrence of PC-2 which we hope that is precursor for PC-4 was very sensitive on change of cadmium concentration. Anyway the matrix and biological variability between each plant could not be minimalized and thus we can be assuming but not certainly sure about these trends.

Conclusion

In presented work we optimized the method for determination of six important biological active thiol compounds which are involved in process of metal bonding in plant of flax. We were able to compare amounts of each thiol with applied concentration of cadmium. This method is could be very helpful in biological studies for observation of plant stress on molecular level. Moreower concentration levels of phytochelatins indicate the enzymatic activity of PCS and thus this method could be directly used   for selecting of proper plats for bioremediation.

Acknowledgements

The work has been supported by GA ČR 522/09/0239, MSMT 6215712402 and 1M06030.

References

Clemens S (2001) Molecular mechanisms of plant metal tolerance and homeostasis. Planta 212: 475-486

Cobbett C, goldsbrough P (2002) Phytochelatins and metallothioneins: Roles in heavy metal detoxification and homeostasis. Annual Review of Plant Biology 53:159-182

Cobbett CS (1999) A family of phytochelatin synthase genes from plant, fungal and animal species. Trends Plant Sci. 4:335-337.

Cobbett CS (2000a) Phytochelatin biosynthesis and function in heavy-metal detoxification. Current Opinion in Plant Biology 3: 211-216.

Cobbett CS (2000b) Phytochelatins and their roles in heavy metal detoxification. Plant Physiol. 123: 825-832

Gawel JE, Ahner BA, Friedland AJ, Morel FMM (1996) Role for heavy metals in forest decline indicated by phytochelatin measurements. Nature 381:64-65

Gawel JE, Trick CG, Morel FMM (2001) Phytochelatins are bioindicators of atmospheric metal exposure via direct foliar uptake in trees near Sudbury, Ontario Canada. Environ. Sci. Technol. 35:2108-2113

Hall JL (2002) Cellular mechanisms for heavy metal detoxification and tolerance. J Exp Bot 53:1-11

Vatamaniuk OK, Bucher EA, Ward JT Rea PA (2001) A new pathway for heavy metal detoxification in animals - Phytochelatin synthase is required for cadmium tolerance in Caenorhabditis elegans. J. Biol. Chem. 276: 20817-20820

Zenk MH (1996) Heavy metal detoxification in higher plants - A review. Gene 179:21-30

J Biochem Tech (2010) 2(5):S13-S15
ISSN: 0974-2328

Electrochemical preparation of polypyrrole conducting films

Mária Filkusová*, Renáta Oriňáková

Received: 25 October 2010 / Received in revised form: 13 August 2011, Accepted: 25 August 2011, Published: 25 October 2011
© Sevas Educational Society 2011

Abstract

Cyclic voltammetry has been used to investigate the electrochemical polymerization of pyrrole on the surface of a paraffin impregnated graphite electrode (PIGE). Effect of pH and concentration of the electrolyte solution on the electrochemical deposition of polypyrrole (PPy) was studied. The structure of the deposited layers was studied using scanning electron microscope (SEM). Well–adhering black PPy films were obtained.

Keywords: pyrrole, polypyrrole, electrochemical polymerization, paraffin impregnated graphite electrode

Introduction

Polypyrrole (PPy) is one of the most promising polymers for technological and biomedical applications [1]. PPy is one of the most widely used polymers in the research and in the industry, where its use is found in antistatic coating, cell culture substrate, flexible electronics, gas and chemical sensors [2], design of biosensors based on immobilized enzymes, antibodies or DNA [3], batteries and corrosion protection coating [2]. In this paper, we studied the influence of electrochemical polymerization conditions such as electrolyte pH and concentration on the morphology and quality of PPy layers.

Materials and methods

All chemicals were of analytical grade or better quality. Distilled water and freshly prepared solutions were used throughout. Pyrrole (Py) monomer (98+%) was obtained from Sigma–Aldrich. All experiments were carried out at room temperature.

Mária Filkusová*

Department of Physical Chemistry, Faculty of Science, Comenius University, Mlynská Dolina, SK–842 15 Bratislava 4, Slovak Republic

*Tel: +421 55 234 2328, Fax: +421 55 622 2124,
E–mail: maria.filkusova@gmail.com

Renáta Oriňáková

Department of Physical Chemistry, Faculty of Science, P.J. Šafárik University, Moyzesova 11, SK–04154 Košice, Slovak Republic

The electrochemical measurements were performed in a classical three–electrode electrochemical cell. The working electrode was a paraffin impregnated graphite electrode (PIGE) composite rod with a 6 mm diameter. The reference electrode was Ag/AgCl/3 mol/l KCl. A large-area platinum electrode served as the counter electrode. To obtain reproducible results, the surface of PIGE was mechanically renewed with emery paper, polished on glossy paper, then washed with distilled water and degreased with acetone. All electrochemical polymerization procedures and electrochemical experiments were performed using an EcaStat potentiostat/galvanostat, model 110 V (Istran, Slovak Republic).
The polypyrrole films were obtained in previously degassed aqueous solution containing 0.6mol/l, 0.06mol/l or 0.006mol/l pyrrole (Py) and 0.1mol/l NaCl at different pH (2.2, 5.6 and 11.2) by cyclic voltammetry (CV) between –800 mV and +1100 mV (vs. Ag/AgCl/3 mol/l KCl), beginning at -800 mV.

Scanning electron microscope (SEM) images were obtained with a TESLA BS 340. The operating voltage for the SEM was maintained at 20 kV throughout the analysis.

Results and discussion

CV Study

The PPy films were prepared by oxidation of the Py monomer on a PIGE. A typical cyclic voltammograms for different Py concentration at pH = 2.2; 5.6; 11.2 are shown in Fig. 1a, b, c. The highest cathodic and anodic currents were observed for highest Py concentration (0.6 mol/l) and lowest pH (2.2) (Fig 1a). The anodic current appearing in the first scan is associated with irreversible oxidation of the pyrrole monomer to produce the Py radical cation. From the second scan a new anodic current appears at gradually less positive potentials. On reverse scans a cathodic wave corresponding to reduction of the oxidized polymer was registered as a counterpart to the reversible anodic current. The overlapping of both oxidation processes in one broad anodic wave was observed for the highest Py concentration. The magnitude of the PPy redox currents increased and shifted to more positive potentials with the voltammetric cycle number indicating the build–up of electroactive PPy film.

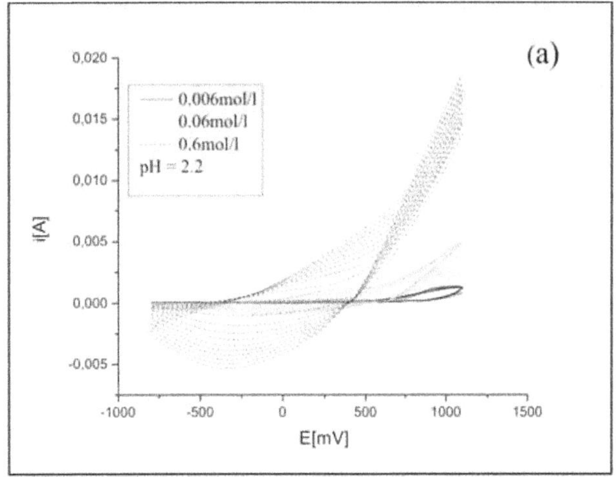
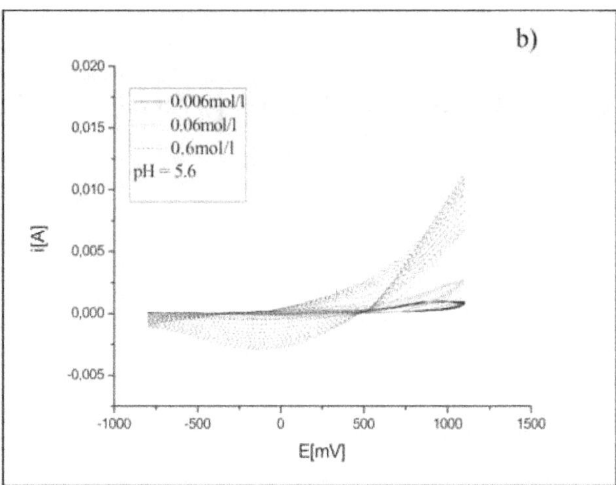
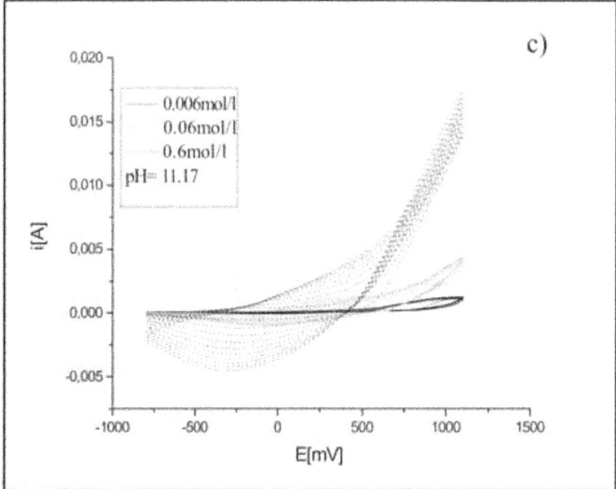

Figure 1: CV curves on PIGE electrode in 0.1 mol/l NaCl electrolyte for different concentration of Py at pH = 2.2 (a); pH=5.6 (b) and pH=11.2 (c)

From the CV studies it can be concluded that the low pH and high Py concentration favored the electropolymerization of Py. High concentration of hydrogen protons is assumed to facilitate the polymerization reaction

SEM Study

The properties of conducting polymers are strongly dependent on their morphology and structure. Scanning electron microscopy was used to investigate the   morphology   of   the   PPy   films electrodeposited. Microscopic observation of PIGE surface after PPy deposition from alkaline media at lower Py concentration showed that the PIGE surface was not completely covered by the polymer. In acidic (Fig 2) and weakly acidic electrolyte, the PPy films present a cauliflower–like structure constituted by micro–spherical grains. In these cases, the electrode becomes rapidly covered by a black, homogeneous, thick and very adherent film. The cauliflower   structure   increased   with   an   increased   in   Py concentration.

Conclusion

The use of an acidic electrolyte solution for the polypyrrole synthesis results in an enhanced electropolymerization and provides a high quality PPy film with potential application in biosensors.

Acknowledgment

The authors wish to acknowledge the financial support from the Grant Agency of Ministry of Education of the Slovak Republic (Grant No. 1/0011/11).

References

Ramanaviciene A, Ramanavicius A (2004) Towards the hybrid biosensors based on biocompatible conducting polymers, in Uv Solid-State Light Emitters and Detectors, Vol. 144. (Nato Science Series, Series Ii: Mathematics, Physics and Chemistry). Edited by Shur MS, Zukauskas A, pp. 287-296

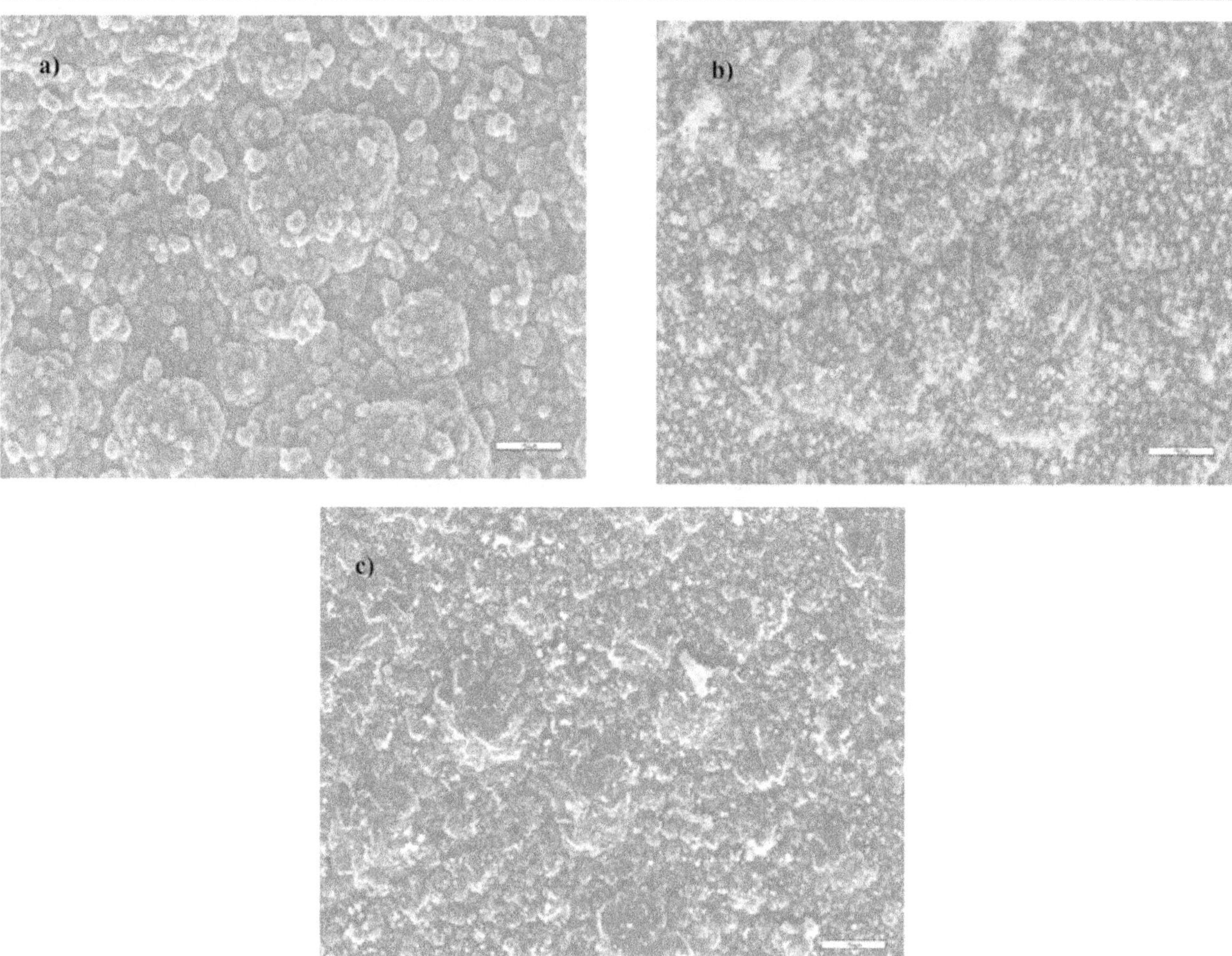

Figure 2: SEM image of PPy film deposited from the electrolyte with concentration 0.6 mol/l (a) 0.06 mol/l (b) 0.006 mol/l (c) at pH = 2.2

Teh KS, Takahashi Y, Yao ZH, Lu YW (2009) Influence of redox-induced restructuring of polypyrrole on its surface morphology and wettability. Sensors and Actuators a-Physical 155(1):113-119

Ramanavicius A, Kausaite A, Ramanaviciene A (2005) Polypyrrole-coated glucose oxidase nanoparticles for biosensor design. Sensors and Actuators B-Chemical 111(532-539

J Biochem Tech (2010) 2(5):S16-S17
ISSN: 0974-2328

Sensitive voltammetric detection of allopurinol–based drug Milurit in clinical urine samples

Stanislav Hason*, Sona Stepankova

Received: 25 October 2010 / Received in revised form: 13 August 2011, Accepted: 25 August 2011, Published: 25 October 2011
© Sevas Educational Society 2011

Abstract

Here we summarize three electrochemical approaches to improve the sensitivity and specificity of the voltammetric sensing of purine metabolites related to the xanthine oxidase (XO) pathway (hypoxanthine-Hyp, xanthine-Xan, and uric acid-UA), as well as of their analogues used as therapeutics in the inhibition of XO enzymatic activity (allopurinol-Alo and its enzymatic product oxypurinol-Oxy) on various carbon-based materials. Detection of purine metabolites, in all proposed protocols, is based on direct electrochemical measurement of oxidation peaks for each of the monitored substances in a single detection step by the same electrode system. A nanomolar detection of these purine metabolites is possible due to: (i) mechanical roughening of the surfaces of glassy carbon or edge plane-oriented pyrolytic graphite electrodes by 15-µm silicon carbide particles (with detection limits around 20 nM), (ii) electrochemical activation of polished carbon-based materials by cycling the potential in 0.1 M KNO_3 between -0.1 and +1.85 V, and (iii) anodic stripping of the electrochemically accumulated purine-Cu(I) complexes from commonly used carbon/graphite materials in the presence of copper ions. Contrary to the mechanical roughening, both electrochemical anodization of carbon-based surfaces and anodic stripping of accumulated purine-Cu(I) complexes are less sensitive to detect Alo in the presence of Hyp due to partial overlapping of their oxidation peaks. However, all of the proposed protocols are operational to fast, sensitive, and inexpensive determination of UA, Xan as well as Oxy in a 10-µl volume of 1000 times diluted urine samples from patients treated with the Alo-based drug Milurit.

Keywords: voltammetry, carbon materials, purine catabolites, urine

---

Stanislav Hason*

Institute of Biophysics, AS CR, v.v.i., Kralovopolska 135, CZ–612 65 Brno, Czech Republic

*Tel.: +420 541 517 261, fax: +420 541 211 293
E-mail: hasons@ibp.cz

Sona Stepankova

Centre for Cardiovascular and Transplant Surgery Brno, Pekarska 53,CZ-656 91 Brno, Czech Republic

Introduction

It is important to monitor the concentration of purine metabolites in cell or body fluids to find out metabolic defects which are characterized by abnormal concentrations of the metabolites resulting in indications of pathological conditions such as gout, hyperuricaemia, Lesch-Nyan disease, renal failure, diabetes, high blood pressure, kidney disease, and heart disease (Lakhsmi et al., 2011). Our previous work has shown that differential pulse voltammetry in connection with mechanically roughened edge plane-oriented pyrolytic graphite (g-PGEe) can serve as a simple and efficient tool for monitoring transformation of purine catabolites (Hyp, Xan, and UA) catalyzed by XO as well as inhibition of this pathway by Alo being enzymatically converted to Oxy (Hason et al., 2009). In addition, mechanically roughened glassy carbon and/or edge plane-oriented pyrolytic graphite electrodes allow simultaneous detection of excreted metabolites related to the XO pathway (including relevant therapeutics) in a 1000 times diluted urine of patients treated or not treated with an Alo-based drug Milurit. This work is aimed at developing the simplest possible electrochemical approaches for improving the sensitivity and selectivity of the voltammetric monitoring of excreted purine metabolites in microlitre volumes of clinical urine samples using carbon-based materials. Stirring of the solution in the accumulation step is an important factor which contributes to amplify the oxidation signals of purine-based molecules. Continuous motion (rotation) of a 10-µl droplet of the analysed diluted urine samples was attained by inert gas streaming (bubbling) through the sample drop, which forms the recently proposed three-electrode inverted drop microcell (Hason et al. 2006; Hason et al. 2008; Hason et al. 2009a,2009b; Lakshmi et al. 2011).

Materials and Methods

Voltammetric measurements were performed using an Autolab PGStat12 potentiostat (Metrohm Autolab B.V., Utrecht, The Netherlands) connected to a three-electrode system involving different carbon/graphite as working electrodes, Ag|AgCl|3MKCl reference electrode, and a platinum wire (1-mm diameter) auxiliary electrode. All measurements were carried out at room temperature either in the classical electrochemical cell (usually in 1 ml volume) or in a homemade inverted drop microcell (usually 10 µl volume). Surfaces of an edge plane-oriented pyrolytic graphite electrode (PGEe) and glassy carbon electrode (GCE) were mechanically roughened (g-PGEe and g-GCE) with silicon carbide paper of 1200-grit (corresponding to 15-μm abrasion particles). The GCE surface was polished (p-GCE) with 1-μm diamond paste. We also performed an electrochemical pretreatment of polished GCE (ea-GCE). The electrochemical activation was conducted with linear cyclic sweep at 50 mV s^{-1} between potentials of -0.1 V and +1.85 V in 0.1 M KNO_3 during 10 cycles. Oxidation peaks of the monitored substances were measured by means of anodic stripping differential pulse voltammetry in 0.1 M acetate buffer (pH 4.8) with the following settings: pulse amplitude of 25 mV; pulse width of 50 ms; scan rate of 15 mV s^{-1}; potential of accumulation (E_A =+0.2 V); accumulation time (t_A = 2 min); rate of stirring of the bulk solution (ω = 3000 min^{-1}); solution in the inverted drop microcell was bubbled by argon at a constant pressure of 0.04 bar. Urine samples from healthy volunteers (not dosed with Alo) and from patients suffering from gout (dosed with 100 mg or 300 mg of Milurit per day for 21 days) was collected in the Department of Internal Medicine and Hepatogastroenterology, University Hospital and Faculty of Medicine, Masaryk University.

Results and Discussion

Fig 1A shows that an equimolar mixture of purine metabolites UA, Xan, Hyp and their structural analogues Alo and Oxy (isosters of Hyp and Xan) produced well-developed and separated voltammetric signals at nanomolar concentrations at the g-PGEe. The inset of Fig 1A shows that the oxidation signals of any of these substances at the g-PGEe were not affected by any significant background current perturbations even at such low concentrations close to the detection limit. Figure 1B compares the current densities of the oxidation peaks of UA, Xan, Oxy, Hyp, and Alo (500 nM each) measured at the g-PGEe, ea-GCE, and p-GCE with a 200 μM Cu(II).

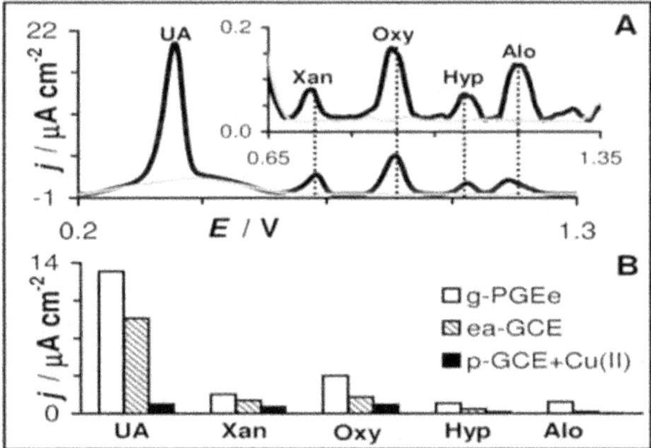

Figure 1 (A) Baseline-corrected differential pulse voltammograms (DPVs) measured for a mixture of UA + Xan + Oxy + Hyp + Alo (500 nM each). The inset shows sections of baseline-corrected DPVs measured for a mixture of UA + Xan + Oxy + Hyp + Alo close to their detection limits (50 nM each). Gray lines represent the blank electrolyte. All curves were recorded at g-PGEe. (B) Current densities of the oxidation peaks of UA, Xan, Oxy, Hyp, and Alo produced by an equimolar mixture of these five substances (500 nM each) measured at different pretreated carbon electrodes (as indicated in the Figure, for details see Experimental). The measurements were performed in the usual 1-ml voltammetric cell.

It can be seen from Fig 2 that DPV response of a 1000 times diluted urine from a healthy volunteer shows only two well-developed and separated oxidation peaks attributable to oxidation of UA (at + 0.42 V) and Xan (at +0.75 V). In the case of patients treated with the drug Milurit containing Alo as the active substance, in addition to the voltammetric peaks of UA and Xan, we can observe a new well-

developed oxidation peak at +0.95 V corresponding to the electrochemical oxidation of Oxy (enzymatic product of Alo).

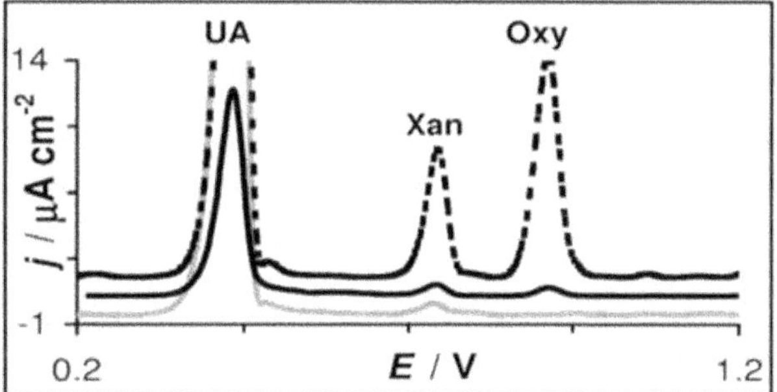

Figure 2 Baseline-corrected DPVs measured at g-PGEe for control human urine (gray curve) and for two human urines collected from patients suffering from gout and dosed with 100 mg (black full curve) and 300 mg (black dashed curve) Milurit per day for 21 days. All urine samples were diluted 1000-fold. The measurements were performed in the inverted 10-μl droplet microcell.

The use of inert gas bubbling to drive the motion of microlitre volumes (10-μl) led to a similar enhancement of oxidation signals of purine metabolites with at least 100-times-reduced sample volume compared to those obtained after accumulation from a stirred solution in the usual 1-ml electrolysis cell (Fig 2).

Conclusion

The electrochemical methods proposed show that urine excretion of the purine metabolites (UA and Xan) as well as of Oxy in patients treated with the Alo–based drug Milurit can easily be monitored electrochemically in a 10-μl drop of the only diluted urine sample in the presence of other drugs such as: antibiotics, blood pressure affecting drugs, antidiabetics, or diuretics.

Acknowledgement

This work was supported by the Grant Agency of the Czech Republic (P205/10/2378), the Academy of Sciences of the Czech Republic (KAN200040651), and institutional research plans (AV0Z50040507, AV0Z50040702).

References

Hason S, Vetterl V (2006) Amplified Oligonucleotide Sensing in Microliter Volumes Containing Copper Ions by Solution Streaming. Analytical Chemistry 78(14):5179-5183

Hason S, Pivonkova H, Vetterl V, Fojta M (2008) Label-Free Sequence-Specific DNA Sensing Using Copper-Enhanced Anodic Stripping of Purine Bases at Boron-Doped Diamond Electrodes, Analytical Chemistry 80(7):2391-2399

Hason S, Stepankova S, Kourilova A, Vetterl V, Lata J, et al. (2009a) Simultaneous Electrochemical Monitoring of Metabolites Related to the Xanthine Oxidase Pathway Using a Grinded Carbon Electrode. Analytical Chemistry 81(11):4302-4307

Hason S, Vetterl V, Jelen F, Fojta M (2009b) Improved Sensitivity and Selectivity of Uric Acid Voltammetric Sensing with Mechanically Grinded Carbon/Graphite Electrodes. Electrochimica Acta 54(6):1864-1873

Lakshmi D, Whitcombe MJ, Davis F, Sharma PS, Prasad BB (2011) Electrochemical Detection of Uric Acid in Mixed and Clinical Samples: A Review. Electroanalysis 23(2):305-320

J Biochem Tech (2010) 2(5):S18-S19
ISSN: 0974-2328

Monitoring of DNA-doxorubicin interactions by electrochemical methods

Hana Kynclova, Dalibor Huska, Jaromir Hubalek, Petr Babula, Tomas Eckschlager, Marie Stiborova, Rene Kizek

Received: 25 October 2010 / Received in revised form: 13 August 2011, Accepted: 25 August 2011, Published: 25 October 2011
© Sevas Educational Society 2011

Abstract

Cancer is presently one of the most studied lifestyle disease. Its treatment is among others based on chemotherapy. One of the drugs used in chemotherapy is doxorubicin, which interacts with DNA and inhibits division of tumour cells. For better usage of this compound as drug, it is necessary to know its interactions with DNA. In this present study interactions between doxorubicin and DNA were monitored using electrochemical methods. In addition, intercalation of doxorubicin into DNA isolated from resistant as well as sensitive tumour cells was monitored and mutual correlation was searched.

Keywords: Cancer, Doxorubcin, DNA, Electrochemical

Hana Kynclova, Jaromir Hubalek

Department of Microelectronics, Faculty of Electrical Engineering and Communication, Brno University of Technology, Udolni 53, CZ-602 00 Brno, Czech Republic

Rene Kizek*

Department of Chemistry and Biochemistry, Mendel University in Brno, Zemedelska 1, CZ-613 00 Brno, Czech Republic

Dalibor Huska

Department of Plant Biology Mendel University in Brno, Zemedelska 1, CZ-613 00 Brno, Czech Republic

Petr Babula

Department of Natural Drugs, Faculty of Pharmacy, University of Veterinary and Pharmaceutical Sciences, Palackeho 1-3, CZ-612 42 Brno, Czech Republic

Tomas Eckschlager

Department of Paediatric Hematology 2nd Faculty of Medicine, Charles University, V Uvalu 84, CZ-150 06 Prague 5, Czech Republic

Marie Stiborova

Department of Biochemistry, Faculty of Science, Charles University, Albertov 2030, CZ-128 40 Prague 2, Czech Republic

*Tel: +420 545 133 350, Fax: +420 545 212 044
E-mail: kizek@sci.muni.cz

Introduction

Recently, interest in development of biosensor methods, which would improve and facilitate investigation and analysis of properties of deoxyribonucleic acid (DNA) and its interactions with other biomolecules, are significantly increasing. It is well known that electrochemical methods are very efficient for these purposes. In addition, these methods can be used for study of cell processes connected with presumptive drug resistance.

Chemotherapy belongs to the group of basic therapeutic strategies for treatment of tumour diseases. Doxorubicin is one of the most used compounds in tumour therapy. Doxorubicin is included in group of anthracycline antibiotics (Huska et al. 2009; Drummond et al. 2003). Its mechanism of action is still unclear, despite of its utilization in therapy. Primary target of anthracyclines is DNA One of the possibilities of action is intercalation, eventually formation of covalent adducts of doxorubicin with DNA nucleotides, which lead to inhibition of processes of replication, eventually transcription.

For better utilization of this drug, it is necessary to study and know its interaction with DNA. In this work, this interaction is studied by the help electrochemical methods. By the use of electrochemical methods, covalent adducts of doxorubicin with DNA isolated from resistant and sensitive cells were studied. Mutual correlation was also searched (Huska et al. 2009; Drummond et al. 2003).

Materials and methods

Electrochemical determination was carried out by the help of apparatus AUTOLAB analyzer (EcoChemie, Netherlands) in connection with VA-Stand 663 (Metrohm, Switzerland). Cell lines were obtained by isolation and selection from neuroblastoma tissue cell cultures. For measurement, DNA isolated from cells sensitive and resistant to doxorubicin was used.

Cell characteristics was measured using sensitive and resistant UKF-NB-4 lines to the presence of cisplatin. DNA was isolated using Wizard Genomic DNA Purification Kit (Promega, Madison, WI USA). The principle is cell lysis, enzymatic digestion of RNA, deproteination and precipitation of genomic DNA by isopropanol. DNA concentration was determined spectrometrically (Hermo spectronic, Orchester NY, USA) at 260 nm. Moreover, the absorbance was measured at 280 nm due to the presence of impurities in the form of proteins (Frederick et al. 1990).

Results and discussion

The statistical analysis was done for comparing the differences between signals of DNA isolated from sensitive and resistant tumour cells. For analysis itself, cyclic voltammetry with evaluation of CA signal was used. From dependencies of signals heights on rates of polarization (0.1, 0.2, 0.3, 0.4, 0.5, 0.6, 0.7, 0.8, 0.9, 1.0 V/s), we determined angular coefficients, which were compared with angular coefficient of control sample (without doxorubicin addition) In Fig 1, value of angular coefficient in dependence on concentration of doxorubicin added into cultivation medium is demonstrated. Differences between angular coefficients of sensitive and resistant tumour cell lines are statistically significant (significance level α=0.05) and average difference of values of angular coefficients of sensitive and resistant cells is ?k= 0,167 ± 0,083. As it follows from experimental data, selection of biochemical processes leading to resistance origination can be expected.

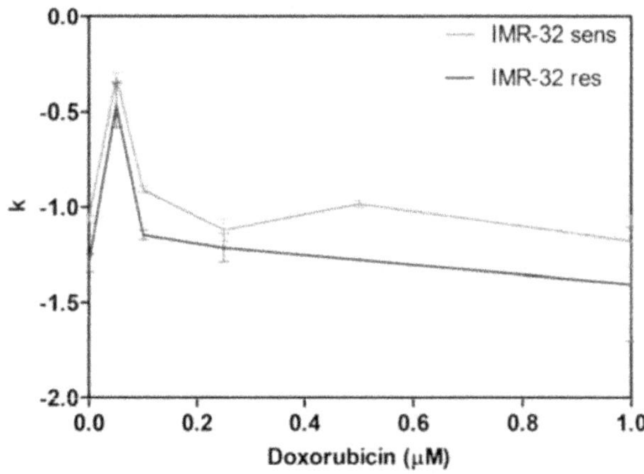

Figure 1: Dependence of values of angular coefficients on doxorubicin concentration

Conclusion

In our experimental work, differences between sensitive and resistant neuroblastoma cell lines were monitored by electrochemical techniques. Doxorubicin proved intercalation, but also covalent bond into DNA.

Acknowledgements

The work has been supported by grants: NANSEMED GA AV KAN20813081, NANIMEL GACR 102/08/1546, GACR P301/10/0356, GA AV IAA40199071 and SIX CZ.1.05/2.1.00/03.0072.

References

Frederick CA, Williams LD, Ughetto G, et al (1990) Structural comparison of anticancer drug-DNA complexes: adriamycin and daunomycin. Biochemistry 29(10): 2538–49.

Drummond,TG, Hill MG, Barton JK (2003) Electrochemical DNA sensors, Nature biotechnology. roč. 21(10):1192-1199

Huska D, Adam V, Babula P et al (2009) Square-Wave Voltammetry as a Tool for Investigation of Doxorubicin Interaction with DNA Isolated from Neuroblastoma Cells, Elektroanalysis, 21, No.3-5, 487-494.

J Biochem Tech (2010) 2(5):S20-S21
ISSN: 0974-2328

Amperometric determination of rutin on carbon paste electrodes

Pavla Macikova, Jan Hrbac, Vladimir Halouzka, Jana Skopalova

Received: 25 October 2010 / Received in revised form: 13 August 2011, Accepted: 25 August 2011, Published: 25 October 2011
© Sevas Educational Society 2011

Abstract

Three different carbon paste electrodes (i.e. unmodified, iron phthalocyanine and ionic liquid modified) were tested to determine rutin by amperometry. The widest linear concentration range and lowest detection limit were obtained with unmodified carbon paste electrode.

Keywords: Rutin, carbon paste electrodes, pulse amperometric technique

Introduction

Rutin (Fig 1) is a bioactive flavonoid, naturally occurring in plants and it is generally present in the common human diet. Numerous analytical methods have been applied to determine rutin, such as voltammetric techniques (Ghica and Brett 2005), capillary electrophoresis (Chen et al. 2000), HPLC (Liu et al. 2008), spectrophotometry (Hassan et al. 1999) and chemiluminescence (Song and Hou 2002).

Electrochemical behavior of rutin is characterized by two oxidative signals. The first reversible anodic signal corresponds to two-electron oxidation of -OH groups at positions 3' and 4' forming an *o*-quinone. The second, irreversible anodic signal is presumably caused by oxidation on the ring A (Ghica and Brett 2005).

Pavla Macikova

Regional Centre of Advanced Technologies and Materials, Department of Analytical Chemistry, Faculty of Science, Palacky University, 17. listopadu 12, 77146 Olomouc, Czech Republic

Jan Hrbac, Vladimir Halouzka

2 Department of Physical Chemistry, Faculty of Science, Palacky University, 17. listopadu 12, 771 46 Olomouc, Czech Republic

Jana Skopalova

3 Department of Analytical Chemistry, Faculty of Science, Palacky University, 17. listopadu 12, 771 46 Olomouc, Czech Republic

*Tel: +420 585 634 442, Fax: +420 585 634 433
E-mail: jana.skopalova@upol.cz

In the literature, a considerable number of the use of amperometric techniques can be found, such as pulse amperometric detection for separation techniques (Ruiz MA et al. 1999) and various amperometric biosensors (Granero AM et al. 2010). Amperometric measurements with various modified carbon paste electrodes (CPEs) are often used for analysis of biological active substances (Forzani et al. 1997). In this work, an amperometric determination of rutin using different modified CPEs is described for the first time.

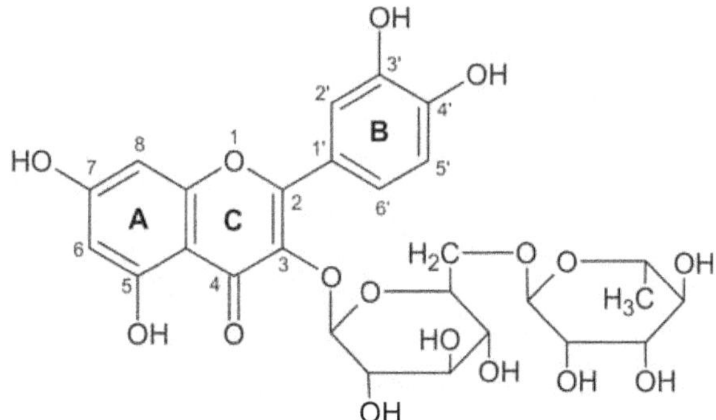

Figure 1: Chemical structure of rutin

Materials and methods

Amperometric measurements in stirred solution were done with CH660 electrochemical workstation (CH Instruments, U.S.A.). The three electrode system involved saturated Ag/AgCl reference electrode, a platinum wire auxiliary electrode and three different carbon paste electrodes as working electrodes, i.e. unmodified CPE, CPE modified with iron phthalocyanine (~ 90 %, Fluka) (IP/CPE) and CPE modified by 1-hexyl-3-methylimidazolium-bis(trifluoromethylsulfonyl)imide [hmim][Tf$_2$N] ionic liquid (⩾98.0 %, Merck) (IL/CPE).

The carbon paste was prepared by mixing 200 mg of graphite flakes (Sigma-Aldrich) with 80 µL of paraffin oil. IP/CPE was prepared by replacement of 10 % (by weight) of graphite flakes by iron (II) phthalocyanine. IL/CPE was prepared from 200 mg graphite and 100 µL [hmim][Tf$_2$N]. Each mixture was homogenized in an agate mortar until a cohesive substance was formed. The paste was filled into the teflon electrode body equipped with a piston (inner diameter 2 mm).

The measurements were carried out in acetate buffer pH=4 prepared from acetic acid (0.1 M) and sodium hydroxide (0.2 M) (both Lachema, Czech Republic). The aliquots (50 µL) of 1 mmol/L rutin standard solution (≥94 %, Sigma-Aldrich) were introduced into electrochemical cell containing 20 mL of base electrolyte using a home–made autosampler. A constant potential was set to 600 mV vs. Ag/AgCl.

Results and Discussion

After repeated injections of rutin standard solutions an increased noise in the corresponding amperometric response appeared for all electrodes, which may be caused by adsorption of rutin or its oxidation products on the electrode surface (Fig 2). The highest level of noise even in low rutin concentration of 5 µmol/L exhibited the IL/CPE electrode, which made it unsuitable for rutin analysis. We have found that significant improvement of signal to noise ratio is achieved, if a pulse technique involving cleaning step at –300 mV for 30 s is used. The detection limits using the pulse technique were 0.32 µmol/L for unmodified CPE, somewhat higher for IP/CPE (0.50 µmol/L) and ten times higher detection limit was determined for IL/CPE (3.07 µmol/L). The widest linear concentration range was found for unmodified CPE (0.25–33.8 µmol/L) and the narrowest one was found for IL/CPE (2.5–12.4 µmol/L).

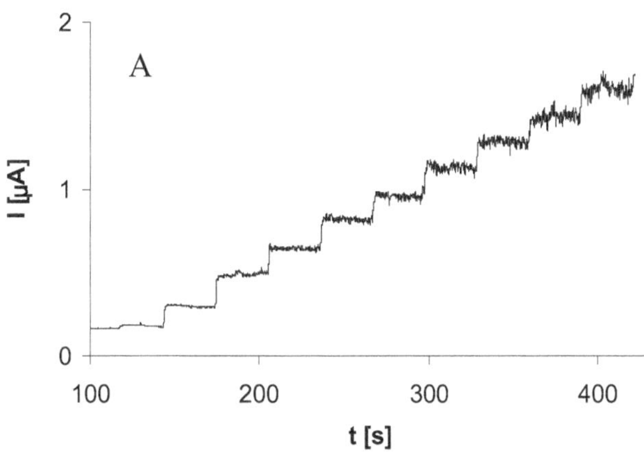

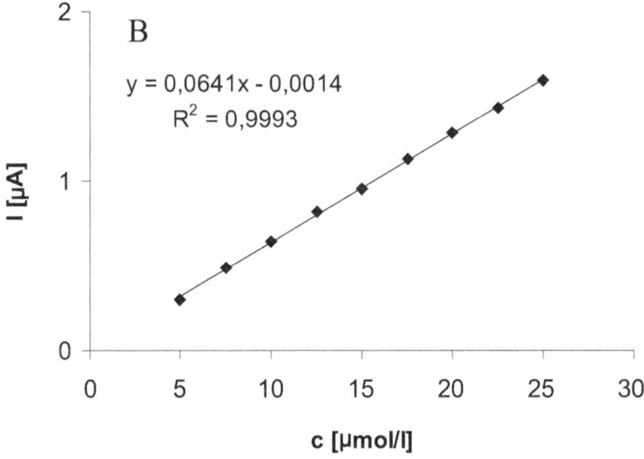

Figure 2: Amperogram (A) and a corresponding calibration curve (B) of rutin. Unmodified CPE, base electrolyte: acetate buffer pH 4, applied potential: 600 mV, stirred solution.

Conclusion

We have developed pulse amperometric technique to determine rutin using carbon paste electrodes.

Acknowledgement

The authors gratefully acknowledge the support by the Ministry of Education, Youth and Sports of the Czech Republic (projects CZ.1.05/2.1.00/03.0058 of Operational Program Research and Development for Innovations - European Social Fund and MSM6198959216). The work has been also supported by the project of Palacky University in Olomouc PRF_2011_025.

References

Chen G, Zhang HW, Ye JN (2000) Determination of rutin and quercetin in plants by capillary electrophoresis with electrochemical detection. Analytica Chimica Acta 423(1):69-76

Forzani ES, Rivas GA, Soils VM (1997) Amperometric determination of dopamine on vegetal-tissue enzymatic electrodes. Analysis of interferents and enzymatic selectivity. Journal of Electroanalyiical Chemistry 435 (1-2):77-84

Ghica ME, Brett AMO (2005) Electrochemical oxidation of rutin. Electroanalysis 17(4):313-318

Granero AM, Fernandez H, Agostini E, Zon MA (2010) An amperometric biosensor based on peroxidases from Brassica napus for the determination of the total polyphenolic content in wine and tea samples. Talanta 83(1):249-255

Hassan HNA, Barsoum BN, Habib IHI (1999) Simultaneous spectrophotometric determination of rutin, quercetin and ascorbic acid in drugs using a Kalman Filter approach. Journal of Pharmaceutical and Biomedical Analysis 20(1-2):315-320

Liu Q, Cai WS, Shao XG (2008) Determination of seven polyphenols in water by high performance liquid chromatography combined with preconcentration. Talanta 77(2):679-683

Ruiz MA, Garcia-Moreno E, Barbas C, Pingarron JM (1999) Determination of phenolic antioxidants by HPLC with amperometric detection at a nickel phthalocyanine polymer modified electrode. Electroanalysis 11(7):470-474

Song ZH, Hou S (2002) Sensitive determination of sub-nanogram amounts of rutin by its inhibition on chemiluminescence with immobilized eagents. Talanta 57(1):59-67

J Biochem Tech (2010) 2(5):S22-S24
ISSN: 0974-2328

A new system for electroanalytical mass screening

Jaromir Zak, Jaromir Hubalek, Rene Kizek*

Received: 25 October 2010 / Received in revised form: 13 August 2011, Accepted: 25 August 2011, Published: 25 October 2011
© Sevas Educational Society 2011

Abstract

The work deals with n–channel system enabling many sample analysis at sort time from sensor array using electrochemical methods. The 8x12 sensor array is formed from 3–electrodes system created on PCB (Printed Circuit Board) with combination of galvanic deposition and screen–printing special paste on the electrodes which can be used for heavy metal analysis and also biomolecular complexes and organic toxic substances determination. The control unit switches each 3–electrode sensor to 8–channel precise potentiostat which was designed for electrochemical analysis with current sensitivity below 1 pA. The potentiostat can synchronize with control unit to depress time for analysis to minimum. Measured data are sent to user–friendly application in computer and analyzed or saved consequently.

Keywords: electrochemical methods, cyclic voltametry, sensor array

Introduction

Nowadays the ecology aspect of human activity is very important because of many toxic substances being produced by industry (Pollard et al., 2009; Yantasee et al., 2007). Rapid monitoring and mapping of toxic contamination of environment is required. The biggest objectives in today's world are to collect a lot of samples, simple and faster measurements by intelligent automatic devices, data archiving and studying saved records. Very promising

Jaromir Zak, Jaromir Hubalek

Laboratory of Microsensors and Nanotechnologies, Department of Microelectronics, Brno University of Technology, Údolní 53, 602 00 Brno

Rene Kizek*

Department of Chemistry and Biochemistry, Mendel University, Zemedelska 1, 602 00 Brno

*Tel: +420 545 133 350, Fax: +420 545 212 044
E-mail: kizek@sci.muni.cz

application can be mass screening in cases of epidemic and pandemic using electrochemical methods (Adam et al., 2009). Time is significant and crucial issue in these cases.

The system design

The whole system consists of a sensor array, a sensor multiplexer, the main device – potentiostat and service program in computer (see Fig 1). Sensor array converts chemical values to electrical signals. There are 96×3 electrodes and their electrical interconnections on the sensor array. These connections are designed for parallel processing. Each electrode has its own connection pin in the connector on the one side of the sensor array, only working electrodes are connected together in each line of sensors. There is shielding layer near the electrodes (ring around and plate under the electrodes) for minimization of external interferences.

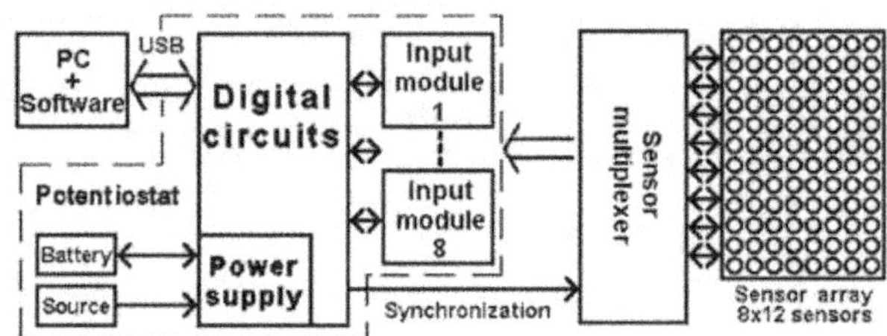

Figure 1: Block diagram of the whole system.

We need a multiplexer system to switch and to group signals to several multiple ones. The multiplexer system is able to select one of the sensors for serial processing or the whole sensor line (group of eight sensors) for parallel processing. The multiplexed signals are distributed to 8-channel potentiostat for real-time multichannel measuring.

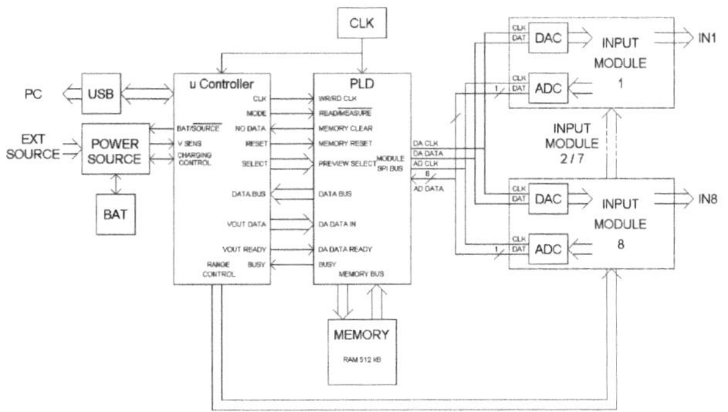

Figure 2: Block schematic of the 8-channel potentiostat (main part of Fig 1).

Simplified schematic of the potentiostat is shown in the Fig 2. The potentiostat is controlled by service application in PC, because standalone measurement would be hard executable (too many measured data would be stored in internal RAM). The service application allows measuring by eight standard electrochemical methods (for example cyclic voltammetry, differential pulse voltammetry, etc.), one user defined method and other methods as EVLS (Elimination Voltammetry with Linear Scan) are being implemented to the application.

Results and Discussion

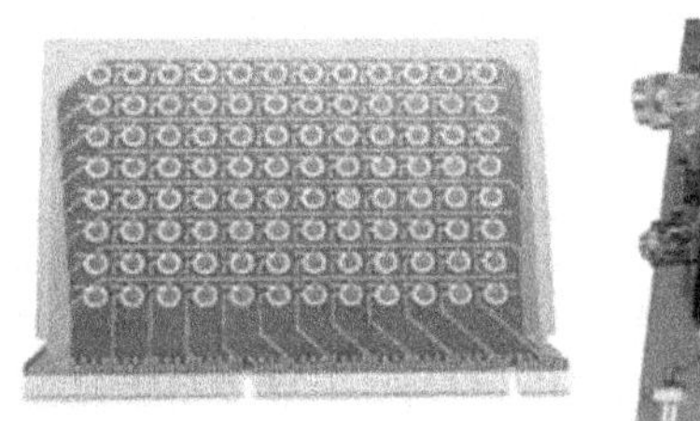 

Figure 3: Sensor array (8 × 12 3–electrode systems) on the left and multiplexer system on the right.

Electrodes were placed to the matrix of sensor array and this array was made on standard organic printed circuit board (PCB, 4-layers FR4) consequently (see Fig 3). Connector pins and sensor electrodes are golden plated (NiAu + galvanic gold). After making basic structure of the sensor array, the auxiliary electrode was deposited from Pt using galvanic method. Working electrode was printed using carbon paste (DuPont, USA), reference electrode was formed by printing using Ag/AgCl paste DuPont, USA). There will be isolating layer around the sensors used for making 3D structure to create holes for specimens. Many different shapes of sensors were made for testing response size and measured specimen stability and two of them were selected and used on sensor arrays (see Fig 4).

Figure 4: Two different shapes of electrodes used on sensor array (red - auxiliary, green – working and yellow – reference electrode)

Multiplexer system can select one column (eight sensors) or each sensor separately. The selection of auxiliary and reference electrodes is realized by analogue CMOS multiplexers (74HC4067 and 74HC4051). Working electrodes are selected by mechanical SMD relays with very low contact resistance. This solution connects maximal accuracy and minimization of the sensor multiplexer size. Sensor multiplexer provides three ways of controlling – manual sensor selection by user, time delayed sensor switching with user defined switching time and external controlling by the superior system, potentiostat for example. Internal multiplexers and relays are switched by AVR RISC microcontroller, actual state is signalized by simple LED display and multiplexer is powered by external stabilized power source. Synchronization link is designed as simple two wires with open collectors and pull-up resistors for maximum versatility (1st wire changes index actual selected sensor and 2nd wire gets sensor to start position).

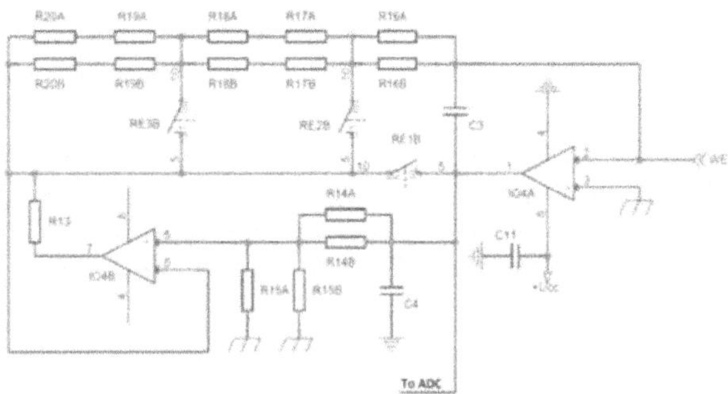

Figure 5: Input module of potentiostat – input analog part schematic

8–channel potentiostat is able to provide more than one measurement at the time which saves the measurement time. The device was build up with SW for controlling the measurement by PC. The potentiostat was designed for parallel high precise/low currents measurement (separated power sources, precise AD/DA converters and analogue circuits, battery powering when the measuring is running with automatic charging, etc.; see Fig 2 and Fig 5). Fig 2 shows modular concept of the potentiostat with one digital controlling block and up to eight input analog modules. To maximize the noise resistance the input modules were designed with separated and shielded blocks (see Fig 6) and their interconnection to the digital part of the potentiostat is realized only by digital bus. Each analog input modul can operate as independent universal measuring system because it contains all necessary parts (power source, programmable voltage output with DAC, voltage input with ADC and I/U converter with five ranges (100 pA, 10 nA, 1 µA, 100 µA and 10 mA).

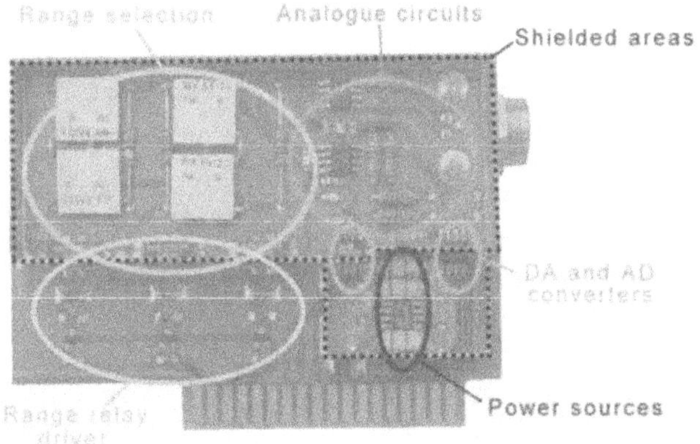

Figure 6: Potentiostat - input analog part HW realisation

The potentiostat measures one row by one row at the time therefore it can synchronize the end of each row measurement with multiplexer to save the time. The instrument is presented in Fig 7. Advanced routines and circuits solutions for increasing resolution and decreasing noise are being implemented.

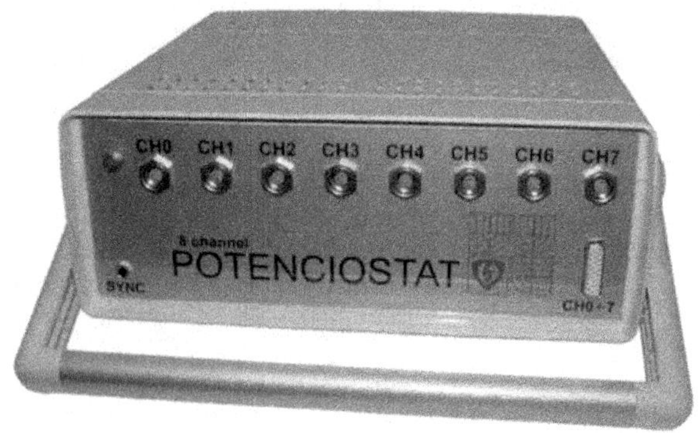

Figure 7: The 8–channel potentiostat

Different concentrations of $PbCl_2$ dissolved in redistilled and deionized water mixed with 4 ml 1 M KCl were used as test analyte. The analyte was applied as drops of 50 µL on electrodes. The measurement method was standard cyclic voltammetry in the range of potentials from 0 V to –1 V with the scan rate of 25 mV/s. There were seven different concentrations measured at the same time. Testing results of Pb ions detection are shown in the Fig 8.

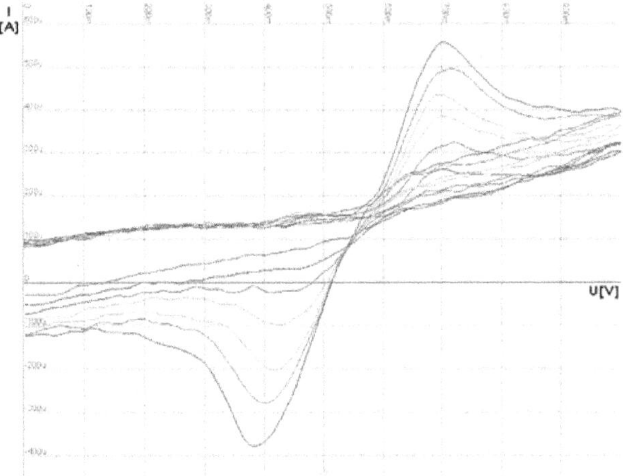

Figure 8: Testing measurement by cyclic voltammetry

Conclusion

This automatic measurement system for mass screening has been developed by accelerating of the precise and rapid electrochemical measurements and increasing their efficiency. Measured data are analyzed by the computer application, specially developed for this device. With the system the mass screening of several substances can be provided during a few minutes. If one measurement takes 2 minutes the 96 samples can be analyzed during 24 minutes.

Acknowledgement

This work has been supported from project GA102/08/1546 and from the frame of research plan MSM 0021630503.

References

Adam V, Zitka O, Huska D, Krizkova S, Strnadel J, et al. (2009) Spatial distribution of heavy metals in healthy and melanoma animal tissues. Febs Journal 276:96-96.

Pollard P, Adams M, Robertson PKJ, Christidis K, Officer S, et al (2009) Environmental Forensic Investigations: The Potential Use of a Novel Heavy Metal Sensor and Novel Taggants: Criminal and Environmental Soil Forensics

Yantasee W, Lin Y, Hongsirikarn K, Fryxell GE, Addleman R, et al. (2007) Electrochemical sensors for the detection of lead and other toxic heavy metals: The next generation of personal exposure biomonitors. Environmental Health Perspectives 115(12):1683-1690

J Biochem Tech (2010) 2(5):S25-S26
ISSN: 0974-2328

Electrochemical study of hepta–oligonucleotides

Zdenka Balcarova, Kamila Neplechova, Libuse Trnkova*

Received: 25 October 2010 / Received in revised form: 13 August 2011, Accepted: 25 August 2011, Published: 25 October 2011
© Sevas Educational Society 2011

Abstract

The study deals with the description and characterization of two hepta–oligonucleotides (DNA and RNA) forming special structures. We studied their electrochemical behaviour by means of cyclic voltammetry (CV) and elimination voltammetry with linear scan (EVLS) in combination with adsorptive stripping (AdS) technique. Differences in electrochemical behaviour of hepta–deoxyribonucleotide and its RNA analog were discussed with regard to their different structures in solutions and their melting points.

Keywords: DNA, RNA, Cyclic voltammetric, electrochemical behaviour

Introduction

Hairpin formation provides a potential structural basis for the constancy of the three–nucleotide regions of several grave diseases. d(GCGAAGC) regions has been found in viral nucleci acid sequences, and in several important prokaryotic and eukaryotic genes (Arai et al. 1981; Cowing et al. 1985; Elias and Lehman 1988; Hirao et al. 1994). This study utilizes electrochemical methods for the study of DNA and RNA heptamers – d(GCGAAGC) and r(GCGAAGC) in aqueous buffered solutions. At mercury electrodes, the hairpin d(GCGAAGC) provides voltammetric reduction overlapped signals of adenine and cytosine and oxidation signals of guanine (Trnkova et al. 2004). The aim of the study was to investigate both types of signals using cyclic voltammetry (CV) and elimination voltammetry with linear scan (EVLS) influenced by various experimental parameters.

Materials and Methods

Synthetic heptanucleotides d(GCGAAGC) and r(GCGAAGC) were synthesized by IDT (Integrated DNA Technology, USA). Buffer components were purchased from Sigma–Aldrich Chemical (purity

Zdenka Balcarova, Kamila Neplechova, Libuse Trnkova*

Masaryk University, Faculty of Science, Chemistry Department, Kotlářská 2, 611 37 Brno

*Tel: +420 549 497 754, Fax: +420 541 211 214
E-mail: libuse@chemi.muni.cz

of ACS). Cyclic voltammetric measurements were carried out with AUTOLAB PGS20 Analyzer (Ecochemie, Netherlands) connected to the VA–stand 663 (Metrohm, Switzerland). The standard cell consisted of three electrodes. Hanging Mercury Drop Electrodes (HMDE) with an area of 0.4 mm^2 were employed as working electrodes. An Ag/AgCl/3M KCl electrode served as the reference electrode. Platinum wire electrode was used as the auxiliary electrode.

Results and Discussion

Both heptamers d(GCGAAGC) a r(GCGAAGC) provide reduction signals of A and C and oxidation signal of G of its reduction product forming at very negative potentials, close to the hydrogen evolution at mercury electrode. These signals were studied by CV and EVLS and the effect of pH, concentration of ODNs, temperature, scan rate, time and potential of accumulation were determined.

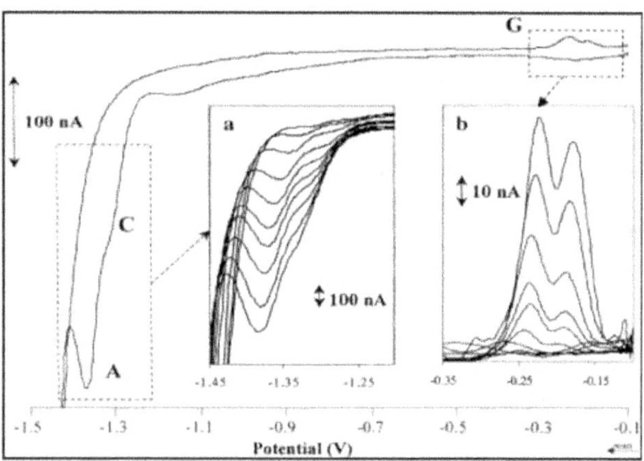

Figure 1: Cyclic voltammograms of d(GCGAAGC) (100nM) at different scan rates (pH 5.35)

Within all pH values, d(GCGAAGC) provides two guanine signals, whereas r(GCGAAGC) only one. Poorly distinguished signals of A and C were resolved by EVLS (Fig 1 & 2).

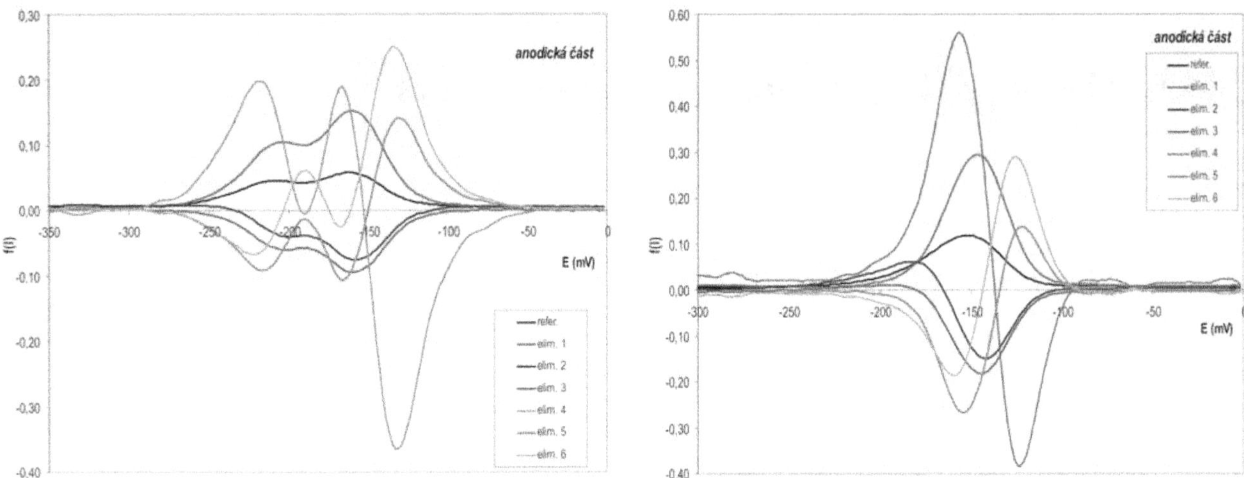

Figure 2: Elimination voltammograms of d(GCGAAGC) and r(GCGAAGC), 1μM; pH 5,3; scan rates 200, 400, 800 mV/s; accumulation at –0.1 V for 60 s

Conclusion

In this study, RNA heptamer was studied to compare it with its DNA analog (Trnkova et al. 2004). It was found that these mini–hairpins differ in their electrochemical behaviour. Replacement of deoxyribose with ribose leads to conformational changes which can be easily detected by electrochemistry.

Acknowledgement

The work has been supported by Ministry of Education (projects MSM 0021622412 and LC06035).

References

Arai K, Low R, Kobori J, Shlomai J, Kornberg A (1981) Mechanism of dnab protein action .5. Association of dnab protein, protein n', and other prepriming proteins in the primosome of dna-replication. Journal of Biological Chemistry 256(10):5273-5280

Cowing DW, Bardwell JCA, Craig EA, Woolford C, Hendrix RW, et al. (1985) Consensus sequence for escherichia-coli heat-shock gene promotersProceedings of the National Academy of Sciences of the United States of America 82(9):2679-2683

Elias P, Lehman IR (1988) Interaction of origin binding-protein with an origin of replication of herpes-simplex virus-1Proceedings of the National Academy of Sciences of the United States of America 85(9):2959-2963

Hirao I, Kawai G, Yoshizawa S, Nishimura Y, Ishido Y, et al. (1994) Most compact hairpin-turn structure exerted by a short dna fragment, d(gcgaagc) in solution - an extraordinarily stable structure resistant to nucleases and heat Nucleic Acids Research 22(4):576-582

Trnkova L, Postbieglova I, Holik M (2004) Electroanalytical determination of d(GCGAAGC) hairpin. Bioelectrochemistry 63(1-2):25-30

J Biochem Tech (2010) 2(5):S27-S28
ISSN: 0974-2328

Possibilities of implementation of elimination voltammetry in electrochemical analyzers

Peter Barath, Jaromír Zak, Libuse Tr nkova*, Jaromir Hubalek, Vojtech Adam, Rene Kizek

Received: 25 October 2010 / Received in revised form: 13 August 2011, Accepted: 25 August 2011, Published: 25 October 2011
© Sevas Educational Society 2011

Abstract

The study is focused on the implementation of elimination voltammetry with linear scan (EVLS) in electrochemical analyzers. This implementation consists in both hardware and software development. First, the hardware, involving the good potentiostat, must enable us to measure voltammetric curves at many different scan rates and with a suitable potential step. Second, the software should be able to provide a program not only for a voltammetric experiment, but also for the elimination procedure. The EVLS procedure needs the smoothing of voltammetric curves, the calculation of coefficients of chosen elimination functions and the presentation of elimination function in both numerical and graphical forms.

Keywords: voltammetry, electrochemical analyzers, potentiostat

Introduction

Thirteen years ago the theory of elimination voltammetry with linear scan (EVLS) was published and experimentally verified for selected electrode systems (Dracka, 1996; Trnkova and Dracka, 1996). To this date this method has found applications not only in electroanalysis, but also in the study of electrode processes of inorganic and organic electroactive substances at mercury, silver or graphite electrodes (Orinakova et al. 2004; Rozik and Trnkova 2006; Trnkova 2002; Trnkova et al. 2000). The EVLS can be considered as a mathematical model of the transformation of current–potential curves capable of eliminating some selected current components, while conserving others by means of elimination functions. For the calculation of the elimination functions three voltammetric curves at different scan rates should be recorded under identical experimental conditions. The attention was devoted to fundamentals of elimination voltammetry and to the transformation of irreversible current–potential curves of an adsorbed electroactive substance. The function, eliminating charging and kinetic current components and conserving the diffusion current component, yields specific, sensitive and well developed peak–counterpeak (p–cp) signal (Adam et al. 2005; Jelen et al. 2009; Mikelova et al. 2007a; Mikelova et al. 2007b; Trnkova 2002; Trnkova and Dracka 1996; Trnkova et al., 2006, 2003; Trnkova et al. 2000; Trnkova et al. 2004; Trnkova et al. 2008).

Materials and Methods

The home–made potentiostat was designed and constructed (Fig 1). New potentiostat has eight inputs for eight voltammetric experiments at the same time. The instrument can measure at five current ranges from $100\,pA$ to $10\,mA$ with smallest measurable value app. $5\,pA$. Internal battery increase noise immunity of the measurement and enable to provide measurement *in situ*. First voltammetric measurements were carried out with the VA–stand 663 (Metrohm, Switzerland). The standard cell consisted of three electrodes, a Hanging Mercury Drop Electrodes (HMDE) with an area of $0.4\,mm^2$ were employed as working electrode, an Ag/AgCl/3M KCl electrode served as the reference electrode and platinum wire electrode was used as the auxiliary electrode.

Results and Discussion

The potentiostat was primarily designed for detecting of various ions and molecules. Using the standard three electrode cell and cyclic voltammetry the first test was devoted to the redox processes of lead (Pb^{2+} in acetate buffer) and the second to the redox system of cadmium (Cd^{2+} in 0.1M KCl).

Peter Barath, Libuse Trnkova*

Department of Chemistry, Faculty of Science, Masaryk University, Kotlářská 2, CZ–611 37 Brno, Czech Republic

Jaromír Zak, Jaromir Hubalek

Department of Microelectronics, Faculty of Electrical Engineering and Communication, Brno University of Technology, Údolní 53, CZ–602 00 Brno, Czech Republic

Vojtech Adam, Rene Kizek

Department of Chemistry and Biochemistry, Faculty of Agronomy, Mendel University in Brno, Zemědělská 1, CZ–613 00 Brno, Czech Republic

*Tel: +420 549 497 754, Fax: +420 541 211 214
E-mail: libuse@chemi.muni.cz

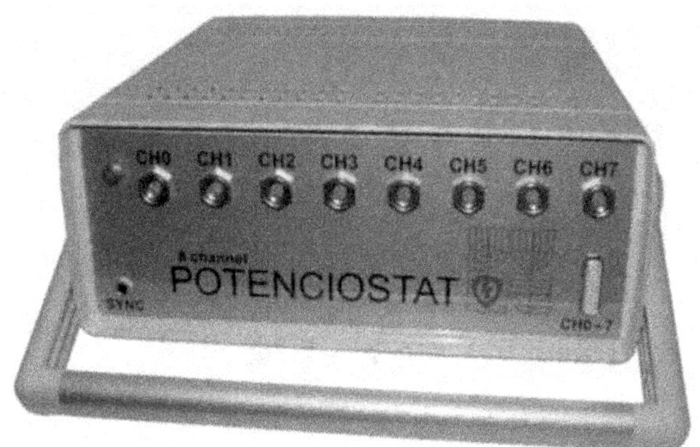

Figure 1: The eight–channel potentiostat. (See page number S22-S25)

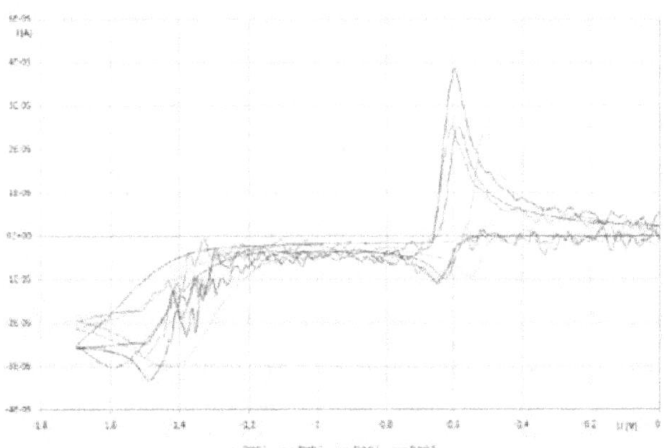

Figure 2: Cyclic voltammograms of redox couple Cd^{2+}/Cd in 0.1M KCl measured with (Autolab – blue lines) and home–made potentiostat (red lines). Scan rate 100 mV/s.

In addition to the measurements of chemical system tests with well–suited instrument inputs were carried out. The results (Fig 2) were discussed and the creation of the suitable filter circuit for suppressing noise in the measurement of real solutions was suggested. Moreover, the curves were successfully treated with elimination voltammetry implemented in the software package.

Conclusion

The small review about the possibilities of the implementation of elimination (EVLS) procedure in electrochemical analyzers was presented. The successful implementation of EVLS will need further the debugging software package and the depression of noise. The smoothing of voltammetric curves, the calculation of coefficients of chosen elimination functions and the graphic presentation of elimination function were designed.

Acknowledgement

This work was supported by the following grants: INCHEMBIOL MSM0021622412, MSM0021630503 (MICROSYN), BIO–ANAL–MED LC06035 from the Ministry of Education, Youth and Sports of the Czech Republic and GAAV grant KAN208130801 (NANOSEMED).

References

Adam V, Petrlova J, Potesil D, Zehnalek J, Sures B, et al. (2005) Study of metallothionein modified electrode surface behavior in the presence of heavy metal ions-biosensor. Electroanalysis 17(18):1649-1657

Dracka O (1996) Theory of current elimination in linear scan voltammetry. Journal of Electroanalytical Chemistry 402(1-2):19-28

Jelen F, Kourilova A, Hason S, Kizek R, Trnkova L (2009) Voltammetric Study of Adenine Complex with Copper on Mercury Electrode. Electroanalysis 21(3-5):439-444

Mikelova R, Trnkova L, Jelen F (2007a) Double elimination voltammetry of short oligonucleotides. Electroanalysis 19(17):1807-1814

Mikelova R, Trnkova L, Jelen F, Adam V, Kizek R (2007b) Resolution of overlapped reduction signals in short hetero-oligonucleotides by elimination voltammetry. Electroanalysis 19(2-3):348-355

Orinakova R, Trnkova L, Galova M, Supicova M (2004) Application of elimination voltammetry in the study of electroplating processes on the graphite electrode. Electrochimica Acta 49(21):3587-3594

Rozik R, Trnkova L (2006) Cadmium reduction process on paraffin impregnated graphite electrode studied by elimination voltammetry with linear scan. Journal of Electroanalytical Chemistry 593(1-2):247-257

Trnkova L (2002) Electrochemical behavior of DNA at a silver electrode studied by cyclic and elimination voltammetry. Talanta 56(5):887-894

Trnkova L, Dracka O (1996) Elimination voltammetry. Experimental verification and extension of theoretical results. Journal of Electroanalytical Chemistry 413(1-2):123-129

Trnkova L, Jelen F, Postbieglova I (2006) Application of elimination voltammetry to the resolution of adenine and cytosine signals in oligonucleotides II. Hetero-oligodeoxynucleotides with different sequences of adenine and cytosine nucleotides. Electroanalysis 18(7):662-669

Trnkova L, Jelen F, Postbieglova I (2003) Application of elimination voltammetry to the resolution of adenine and cytosine signals in oligonucleotides. I. Homooligodeoxynucleotides dA(9) and dC(9). Electroanalysis 15(19):1529-1535

Trnkova L, Kizek R, Dracka O (2000) Application of elimination voltammetry to adsorptive stripping of DNA. Electroanalysis 12(12):905-911

Trnkova L, Postbieglova I, Holik M (2004) Electroanalytical determination of d(GCGAAGC) hairpin. Bioelectrochemistry 63(1-2):25-30

Trnkova L, Zerzankova L, Dycka F, Mikelova R, Jelen F (2008) Study of copper and purine-copper complexes on modified carbon electrodes by cyclic and elimination voltammetry. Sensors 8(1):429-444

J Biochem Tech (2010) 2(5):S29-S30
ISSN: 0974-2328

Determination of metallothionein, glutathione and termostable protein fraction concentrations in tumour tissue from head and neck area

Natalia Cernei, Petr Majzlik, Ondrej Zitka, Dalibor Huska, Hana Binkova,
Michal Masarik, Vojtech Adam, Rene Kizek*

Received: 25 October 2010 / Received in revised form: 13 August 2011, Accepted: 25 August 2011, Published: 25 October 2011
© Sevas Educational Society 2011

Abstract

Tumours of head and neck are one of the most incident tumour diseases. Well-timed diagnosis plays main role in their therapy. Aim of this work consisted in determination of metallothionein (MT), which could be new tumour marker. Determination of its concentration was carried out using electrochemical detection with use of differential pulse voltammetry. Concentrations of MT in samples of denaturized blood plasma of group of patients were about 2.3 µM with slight differences in accordance with localization of tumours. Contents of glutathione and total proteins in denaturized were determined chromatographically and by the help of Biuret reaction. Rate of concentrations levels of GSH/GSSG was varying from 1.31 to 17.66.

Keywords: Metallothionein, GSH and GSSG, new tumour marker, termostable proteins.

Introduction

Annually, many millions of people suffer from different types of tumour disease; unfortunately, effective way of treatment is still

Natalia Cernei, Petr Majzlik, Ondrej Zitka, Dalibor Huska, Vojtech Adam, Rene Kizek*

Department of Chemistry and Biochemistry Faculty of Agronomy, Mendel University in Brno, Zemedelska 1, CZ-613 00 Brno, Czech Republic

Hana Binkova

Department of Otolaryngology and Maxillofacial Surgery, University Hospital, Pekarska 53, CZ-656 91 Brno, Czech Republic

Michal Masarik

Department of Pathological Physiology, Faculty of Medicine, Masaryk University, Komenskeho namesti 2, CZ-662 43 Brno, Czech Republic

*Tel: +420 545 133 350, Fax: +420 545 212 044
E-mail: kizek@sci.muni.cz

missing. In therapeutic procedures, well-timed diagnosis plays crucial role. Metallothionein, which is promptly and exactly determinable, was in our work used as potential marker of tumour disease. Metallothionein is small protein with high content of amino-acid cysteine, without aromatic amino-acids. Metallothionein consists of two domains α and β, which are together able to bind up to seven ions of divalent metals. This protein is excellent transporter of metals, in physiological state especially of zinc and copper. Antioxidant activity of metallothionein was also determined1. Its increased expression in some types of tumour diseases was documented before.

Metallothionein is supposed to participate in conception of resistance of tumour cells to certain types of chemotherapeutics, such as cisplatin. Structural characteristics make metallothionein to be excellent candidate for electrochemical detection. Prof. Brdicka was interested in similar ways of analysis more than 70 years ago. Result of analysis consists in curve with three signals, which the last is catalytic and depends on metallothionein concentration in sample. For analysis, we chose samples of blood from patients suffering by tumours in neck and head area. Moreover we studied a rate of concentrations of glutathione reduced and oxidized (GSH/GSSG) which provides information of oxidative stress. When this rate is less than value of 10 oxidation stress in organism is high. MT and glutathions are thiols which means that they are part of termostable protein fraction of the sample of blood is denaturized (Kagi et al. 1988).

Materials and methods

Samples were analyzed on apparatus 747 VA Stand in connection with 746 VA Trace Analyzer and 695 Autosampler (Metrohm, Switzerland) in classic three-electrode arrangement. Working electrode was hanging mercury drop electrode (HMDE) with drop area 0.4 mm^2, reference electrode was Ag/AgCl/3M KCl and auxiliary was platinum electrode. Basic electrolyte (1 mmol/l $[Co(NH_3)_6]Cl_3$ and 1 mol/l ammonium buffer; $NH_3(aq)$ + NH_4Cl, pH 9.6) was changed for each five measurements. DPV parameters were as follows: initial potential –0.7 V, end potential –1.75 V, modulation time 0.057 s, time interval 0.2 s, potential step 2 mV/s, modulation amplitude -250 mV, Eads = 0 V, temperature of basic electrolyte 4 °C. Samples of blood of patients suffering from tumours in neck and head areas were obtained from Department of Othorhinolaryngology and Head and Neck Surgery of St. Anne´s University Hospital Brno, all with permission of ethic commission. Samples were prepared by following procedure: firstly, they were 100x diluted by phosphate buffer (pH 6.9, 20mM) and subsequently denatured at temperature 99°C for 15 min (Eppendorf 5430, USA). After it, samples were centrifuged (16 000 g, 4 °C, for 30 min; Eppendorf 5402, USA) with subsequent supernatant sampling, where GSH and GSSG were analyzed by liquid chromatography with electrochemical detection using reversed phase chromatographic column Zorbax eclipse AAA C18 (150 × 4.6; 3,5 µm particles, Agilent Technologies, USA) and twelve-channel CoulArray electrochemical detector (Model 5600A, ESA, USA). Mobile phase consists of A: trifluoroacetic acid (80 mM) a B: 100% Met-OH. Detection potential was 900mV. Determination of thermostabile proteins was carried out using Mindray BS 200 and Biuret reaction.

For determination of the total protein content the biuret solution (15 mM potassium sodium tartrate, 100 mM NaI, 15 mM KI and 5 mM $CuSO_4$). As a standard albumin (1 mg in 1 ml of phosphate buffer, pH 7) was used. The measurement was done as follows: 180 µl of the biuret solution was mixed with 45 µl of real or standard sample; after stirring and incubating (10 min. at 37°C) the absorbance at 546 nm was measured.

Results and discussion

Aim of this work was at electrochemical determination of metallothionein. Quantitative information may be obtained in electrochemistry by analysis of two group of electrochemical signals – redox, which connected with electrochemical reaction of analyte, and catalytic. Catalytic signals are connected with reaction of next compounds (for example hydrogen), which originate at given conditions in presence of analyte. Brdicka reaction and H peak technique belong to the group of methods monitoring cathodic signal of hydrogen (Breazeale et al. 1996). These methods are very sensitive and in view of sample preparation very unpretentious. We analyzed 65 samples of blood of patients with malignant tumour in oropharynx localization, 15 samples of full blood of patients with malignant tumour in buccal cavity, 14 samples of full blood of patients suffering from in hypopharynx localization, 5 samples of blood of patients with malignant tumour in paranasal sinuses localization, 29 samples of blood of patients with malignant tumour in larynx localization and 5 samples of blood of patients suffering from malignant tumour of parotid glands. All samples were of first capture. Resulting values of metallothionein concentrations was about 2.3 ± 0.3 µM with slight differences in accordance with localization of tumours (oropharynx 2.8 ± 0.4 µM, buccal cavity 2.5 ± 0.2 µM, hypopharynx 2.2 ± 0.4 µM, paranasal sinuses 2.0 ± 0.3 µM, larynx 2.4 ± 0.5 µM and parotid glands 2.0 ± 0.4 µM). Concentrations were compared with average value of metallothionein concentration in samples of full blood of volunteers (0.5 ± 0.1 µM). Moreover we observed that rate of concentrations levels (GSH/GSSG) was varying from 1.31 to 17.66. That shows the significance differences between each patitent caused probably by different stages of illness and different type of tumors.

In addition to analysis of full blood, levels of MT, total thermostabile proteins and GSH/GSSG were analyzed directly in tumour tissue.

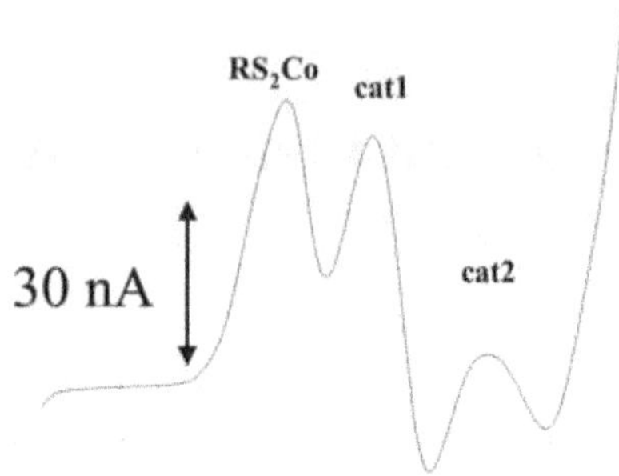

Figure 1: Voltammogram of sample of patients suffering from malignant tumour of larynx. The last signal (Cat2) corresponds to catalytical reduction of hydrogen ions of electrolyte and is directly proportional to concentration of metallothionein in sample.

Conclusion

After comparison of metallothionein levels in patients and controls it is well evident, that metallothionein levels in case of patients are enhanced. By this fact we verified theoretical presumption for using of metallothionein as tumour marker; suitability of electrochemical techniques for its determination and analysis of proteins was also proved. Electrochemical methods would have in proteomics their position, especially thank to their low costingness, rapidity and sensitivity.

Acknowledgements

The work has been supported by grants: NANSEMED GA AV KAN20813081, GA ČR 301/09/P436 and IGA MZ 10200-3

References

Breazeale RI, Fishburn J, Buchanan T, Stone J (1996) Over-expression of Metallothionein and Survival in Uveal Melanoma. Investigative Ophthalmology & Visual Science 37:2877

Kagi JHR, Schaffer AA (1988) Biochemistry of metallothionein.. Biochemistry 27:8509

J Biochem Tech (2010) 2(5):S31-S32
ISSN: 0974-2328

Discovery of two–dimensional condensation of nucleic acids components at the mercury electrodes – 45 years' history

Stanislav Hason*, Vladimir Vetterl

Received: 25 October 2010 / Received in revised form: 13 August 2011, Accepted: 25 August 2011, Published: 25 October 2011
© Sevas Educational Society 2011

Abstract

Purine and pyrimidine derivatives currently occurring in nucleic acids posses an extraordinary high ability of self–association at the electrode surface and can form there by a two–dimensional (2D) condensation a compact self-assembled monolayer. By this high condensation ability nucleic acid bases differ from most of the other purine and pyrimidine derivatives which currently do not occur in nucleic acids. A regular arrangement of nucleic acid bases, nucleosides and nucleotides leading to 2D condensation takes place not only at the liquid mercury surface, but also on the mercury film modified carbon/graphite, solid amalgam, or atomically flat single-crystal metal electrodes. With polymeric DNA and/or polynucleotides this kind of 2D condensation has not been observed. Recently, we have shown that synthetic homo- and hetero-pyrimidinic oligodeoxynucleotides (ODNs) can 2D condensate at the negatively charged mercury and solid amalgam surfaces. Formation of the ODN condensed film took place even in the absence of any ODN in the bulk solution (*ex situ* 2D condensation). No such 2D condensed monolayer was observed with homo- and hetero-purinic ODNs giving only tensammetric (desorption/reorientation) peaks in weakly alkaline pHs within the same potential region.

Keywords: Two-dimensional condensation, adsorption, mercury-based electrodes, adenosine, cytidine, oligodeoxynucleotides

Introduction

The adsorption of nucleic acids at the electrodes can be followed by measurement of the impedance and/or differential capacitance of the electrode double layer, as it was done for the first time in 1961 by Miller (Miller, 1961). In 1965 one of us (VV) discovered that nucleic acid bases as well as some of the nucleosides and nucleotides currently occurring in nucleic acids posses an

Stanislav Hason*, Vladimir Vetterl

Institute of Biophysics, AS CR, v.v.i., Kralovopolska 135, CZ–612 65 Brno, Czech Republic

*Tel: +420 541 517 261, Fax: +420 541 211 293
E-mail: hasons@ibp.cz

extraordinary high ability of self–association at the electrode surface and can form there by a two–dimensional (2D) condensation a monomolecular compact film (Vetterl 1965; Vetterl 1966).

Discussion

The driving forces leading to 2D condensation of nucleic acid bases are hydrogen bonds between flat oriented adsorbed molecules and/or stacking interactions between perpendicularly oriented molecules to the electrode surface. In the presence of the 2D condensed films all solvent molecules are replaced from the electrode surface by the condensed nucleic acids components and thus the differential capacitance of the electrode double layer is depressed considerably, giving rise to a characteristic "pit" on capacitance-potential (*C-E*) curves and sharp spikes on cyclic voltammograms (CVs). The positions of both current spikes and pit edges strongly depend on the direction of the potential scan (hysteresis) as shown in Fig 1, and the width of this condensed film decreases with increasing temperature. These characteristics are

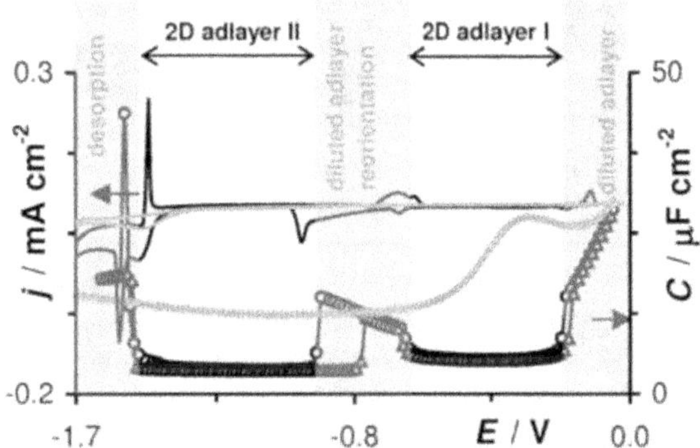

Figure 1: Cyclic voltammogram and capacitance-potential (*C-E*) curves of 14 mM adenosine at pH 5 in 0.1 M NaCl on the hanging mercury drop electrode (HMDE). Potential scan from -0.05 V to negative values ( ) and in the opposite direction (Δ). Gray line and (+) represent the CV and *C-E* responses of background electrolyte, respectively. All measurements were performed at 5° C with scan rate 1 V s^{-1}(CV) and frequency 33 Hz, *ac* voltage amplitude 5 mV, integration time 2 s (*C-E* curves).

typical properties of 2D physisorbed condensed films (Buess-Herman 1994). From a quadratic dependence of the capacitance pit width on temperature and/or from a surface tension measurement was possible to determine the interaction energies of adsorbed molecules and calculate the area occupied per one adsorbed molecule, which gave information about orientation of molecules at the electrode surface (Brabec et al. 1996). With the monomeric nucleic acid bases, the capacitance pits (2D adlayer I in Fig 1) on the mercury electrode were observed usually (with the exemption of cytosine) at potentials close to the potential of zero charge (at about -0.5 V). In contrast, some of the purine and pyrimidine nucleosides formed at the negatively charged surface a second 2D condensed film (2D adlayer II in Fig 1) in which the adsorbed molecules have a different orientation (Hason et al 2002a).

The formation of 2D condensed films on homogenous surface is characterized by the presence of phase transitions. The kinetics of the film formation can be studied by potential jump experiments and often proceed via nucleation and growth processes. Depending on the start and final potentials different shapes of capacitance or current transitions can be detected. The transitions can be analyzed by Avrami theorem (Buess-Herman 1994).

At the beginning of the 1990s, it was observed that the nucleic acid components can form 2D condensed films also on basal faces of single crystal electrodes (Wandlowski 2002). Recently, we have shown that solid amalgam electrodes (thickness of amalgam layer was 40 nm) can be successfully used as a cheaper alternative to very expensive single-crystal metal electrodes for the study of the adsorption, formation of 2D condensed film (for more details see Fig 2) and kinetics of the phase transitions of nucleic acids components (Hason et al. 2004a,b) and adsorption behavior of synthetic ODNs (Hason et al. 2008). 2D condensation can be also studied on the carbon/graphite electrodes modified with a mercury films whose thicknesses are higher than 200 nm (see curve (e) in the Fig 2). In the case of thin (several tens of nanometers) mercury film electrodes was observed that Hg exists on the carbon surfaces in the form of inhomogeneously scattered droplet of different diameter (Hason et al. 2002b,c; Hason et al. 2003) and thus these electrodes have a small sensitivity to detection of the 2D condensation adlayers. In addition, these mercury film electrodes have a low lifetime in comparison with solid nontoxic amalgam electrodes.

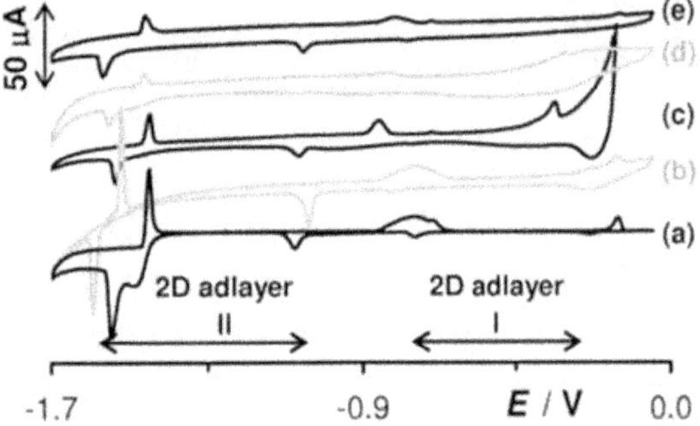

Figure 2: CV curves of 14 mM adenosine at pH 5 in 0.1 M NaCl on the: (a) HMDE, (b) PtAE, (c) CuAE, (d) AgAE, and (e) Hg-PGEb. Two different 2D physisorbed condensed adlayers of adenosine in the acid pH, first of these exists at about -0.5 V, and the second is centered at potential of -1.2 V, are for all mercury hanging drop, solid amalgams, and pyrolytic graphite electrode modified by a mercury film delimited by the sharp current peaks. All measurements were performed at 5° C with scan rate 1 V s^{-1}. The thicknesses of the amalgam layers deposited on the metal substrates and mercury film on the cleaved pyrolytic graphite in basal orientation were 40 nm and 200 nm, respectively.

Conclusion

We have shown that solid amalgam electrodes can serve as an effective tool for study of the adsorption, formation of 2D condensed films and kinetics of the phase transitions of nucleic acids components. We also for the first time demonstrated that synthetic homo- and hetero-pyrimidinic ODNs can 2D condensate at the negatively charged solid amalgam electrodes. Homo- and hetero-purinic ODNs did not form these 2D physisorbed condensed monolayers, only produce tensammetric (desorption/reorientation) peaks in weakly alkaline pHs within the same potential region.

Acknowledgement

This work was supported by the grant projects 202/08/1688 (to V.V.) and P205/10/2378 (to S.H.) of the Grant Agency of the Czech Republic, grant project KAN 200040651 (to S.H.) of the Grant Agency of the Academy of Sciences of the Czech Republic and institutional research plans (AV0Z50040507, AV0Z50040702).

References

Brabec V, Vetterl V, Vrana O (1996) Electroanalysis of biomacromolecules, in: Brabec V, Walz D, Milazzo G (Eds.), Experimental Techniques in Bioelectrochemistry, Birkhauser Verlag, Basel, Swityerland, p.287-359

Buess-Herman C (1994) Self-assembled monolayers at electrode metal surfaces. Progress in Surface Science 46(4) 335-375

Hason S, Vetterl V (2002a) On the formation kinetics of two-dimensional cytidine films. Bioelectrochemistry 57(1):23-32

Hason S, Vetterl V (2002b) Two-dimensional condensation of nucleic acid components at mercury and gold electrodes. Bioelectrochemistry 56(1-2):43-45

Hason S, Vetterl V (2002c) Application of carbon electrodes modified with a mercury layer of a different thickness for studies of the adsorption and kinetics of phase transients of cytidine. Journal of Electroanalytical Chemistry 536(1-2):19-35

Hason S, Simonaho S-P, Silvennoinen R, Vetterl V (2003) On the adsorption and kinetics of phase transients of adenosine at the different carbon electrodes modified with a mercury layer. Electrochimica Acta 48(6):651-668

Hason S, Vetterl V (2004a) Application of thin film mercury electrodes and solid amalgam electrodes in electrochemical analysis of the nucleic acids components:detection of the two-dimensional phase transients of adenosine. Bioelectrochemistry 63(1-2):37-41

Hason S, Simonaho S-P, Silvennoinen R, Vetterl V (2004b) Detection of phase transients in two-dimensional adlayers of adenosine at the solid amalgam electrode. Journal of Electroanalytical Chemistry 568(1):65-77

Hason S, Vetterl V, Fojta M (2008) Two-dimensional condensation of pyrimidine oligonucleotides during their self-assemblies at mercury based surfaces. Electrochimica Acta 53(6):2818-2824

Miller IR (1961) Structure of DNA and RNA in water-mercury interface. Journal of Molecular Biology 3(3):229-240.

Vetterl V (1965) Adsorption of dna components on mercury electrode. Experientia 21(1):9-11

Vetterl V (1966) Differentielle kapazitat der elektrolytischen doppelschicht in anwesenheit einiger purin- und pyrimidinderivate. Collection of Czechoslovak Chemical Communications 31(5):2105-2126

Wandlowski T (2002) Encyclopedia of Electrochemistry, in: Urbakh M, Gileadi M (Eds.), Phase Transitions in Two-Dimensional Adlayers at Electrode Surfaces: Thermodynamics, Kinetics, and Structural, vol. 1, Wiley-VCH, Weinheim, p.383

J Biochem Tech (2010) 2(5):S33-S34
ISSN: 0974-2328

Effect of copper ions on voltammetric signals of aminopurines at a carbon electrode

Frantisek Jelen, Libuse Trnkova, Stanislav Hason, Helena Kudrnova, Vladimir Vetterl, Alena Kourilova

Received: 25 October 2010 / Received in revised form: 13 August 2011, Accepted: 25 August 2011, Published: 25 October 2011
© Sevas Educational Society 2011

Abstract

Electrochemical oxidations of aminopurines (adenine–Ade, 2–aminopurine, 2AP, 2,6–diaminopurine, 2,6–DAP) and their complexes with Cu(I) on carbon electrodes (pencil–PeGE) were investigated by means of linear sweep voltammetry (LSV) and elimination voltammetry with linear scan (EVLS). We found that the Cu(I)–purine complex was formed not only by aminopurines but also by purine. On the other hand the complex is not formed in the case of imidazol or cytosine. The results showed that carbon electrodes in connection with EVLS can be an excellent prototype for cheap and fast working sensor for aminopurines in the presence of copper.

Keywords: Copper, voltammetry, aminopurines

Introduction

Anodic signal of Ade and guanine on the solid carbon electrode was studied and described in details by Dryhurst (Dryhurst 1977). The oxidation involves electrons exchange at carbon C2, C8 and N7 (peak Ox_{Ade}, potential about 1.2 V). Copper ions Cu(I) give an additional oxidation voltammetric peak (peak Ox_{com}) with simultaneous enhancement of peak Ox_{Ade} (Jelen et al. 2004; Shiraishi and Takahashi 1993). The former peak appeared at less positive potential (ca. 0.4 V) due to the re–oxidation of Ade–Cu(I) complex, which had been formed at electrode at sufficient starting potential and accumulation time. In this study, we use simple transfer technique for the electrochemical analysis of Ade and some purine derivatives in the presence of Cu(II) in connection with elimination voltammetry (Trnkova and Dracka 1996, Trnkova 2007).

Materials and Methods

Voltammetric measurements (LSV) were performed with an AUTOLAB analyzer (EcoChemie, The Netherlands) connected with a VA–Stand 663 (Metrohm, Zurich, Switzerland). A standard cell with three electrodes was used. The working electrode was a PeGE with the surface area of 1.67 mm^2 (Tombow 05 HB, Japan). The electrode Ag/AgCl/KCl (3 M) as a reference electrode and platinum wire as an auxiliary electrode were used. Three LSV curves were measured at the same experimental conditions for three scan rates and were taken into EVLS procedure.

Results and Discussion

Using LSV we studied selected aminopurines (Ade, 2-AP and 2,6-DAP) on PeGE . In the absence of Cu(II) ions Ade, 2-AP and 2,6-DAP yielded small oxidative signals. The essential difference at these substances was observed when monovalent copper was generated on the electrode. Aminopurines with Cu(I) form a complex which is responsible for a new oxidative signal (Ox_{Com}) and for enhanced signals of corresponding aminopurines. The Fig. 1a and Fig. 1b show LSV and EVLS results, respectively.

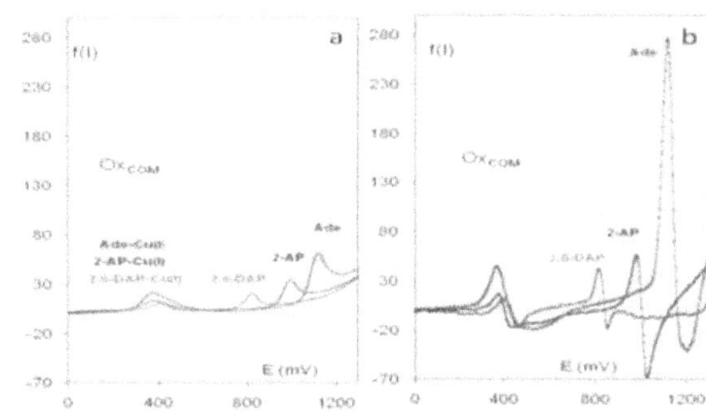

Figure 1: LSV (1a) and EVLS curves (1b) of Ade, 2-AP and 2,6-DAP (10 µM) with 20 µM Cu(II), reference scan rate 250 mV/s

Frantisek Jelen, Stanislav Hason, Helena Kudrnova, Vladimir Vetterl, Alena Kourilova

Institute of Biophysics AS CR, v.v.i., Kralovopolska 135, CZ–612 65, Brno, Czech Republic

Libuse Trnkova

Department of Chemistry, Faculty of Science, Masaryk University, Kotlarska 2, CZ–611 37 Brno, Czech Republic

*Tel: +420 549 497 754, Fax: +420 549 492 443
E-mail: libuse@chemi.muni.cz

LSV in connection of EVLS revealed that in case of Ox_{com} peak, the mechanism of total electrode reaction can be described as an ECE mechanism where electrochemical steps reduction of Cu(II) to Cu(I) and oxidation of Cu(I) from purine–Cu(I) complex. Chemical process represents complexion between Cu(I) and purine with subsequent adsorption of complex on the electrode surface. The complex formation and its oxidation can be described by following scheme:

a)   Cu(II) + e = Cu(I) (at deposition potential –0.15 V)
b)   Cu(I) + purine = [purine– Cu(I)] (on PeGE surface)
c)   [purine– Cu(I)= [purine– Cu(I)]$_{ads}$ (adsorption of the complex)
d)   [purine– Cu(I)]$_{ads}$ = [purine– Cu(II)]$_{ads}$ +e (stripping, peak Ox_{Com})
e)   [purine– Cu(II)]$_{ads}$ = purine$_{ox}$ + Cu(II) +e (stripping, peak Ox).

Conclusion

We found that the studied Cu(I)–Ade complex was formed not only by aminopurines (Ade, AP and DAP) but also by purine. On the other hand the complex is not formed in the case of imidazol and cytosine. According to presented and previous results it can be assumed that the complex formation involves not only nitrogen in the position 9, but also nitrogen atom of pyrimidine ring.

Acknowledgement

This work was supported by the Academy of Sciences of the Czech Republic (grantKAN200040651), the Ministry of Education, Youth and Sports of the Czech Republic (INCHEMBIOL MSM0021622412 and BIO–ANAL–MED LC06035), and institutional research plans (AV0Z50040507, AV0Z50040702).

References

Dryhurst G: Electrochemistry of Biological Molecules. New York: Academic Press, 1977

Jelen F, Kourilova A, Pecinka P, Palecek E (2004) Microanalysis of DNA by stripping transfer voltammetry. Bioelectrochemistry 63(1-2):249-252

Shiraishi H, Takahashi R (1993) Accumulation of Adenine and Guanine as Cu+ Compounds at Glassy-Carbon Electrodes Followed by Anodic-Stripping Vo ltammetry. Bioelectrochem. Bioenerg 31(2):203-213

Trnkova L, Dracka O (1996) Elimination voltammetry. Experimental verification and extension of theoretical results. Journal of Electroanalytical Chemistry 413(1-2):123-129

Trnkova L, in: V. Adam and R. Kizek (Eds.), Utilizing of Bio-electrochemical and Mathematical Methods in Biological Research, Research Signpost, Kerala, India, 51, Ch. 4, 2007

J Biochem Tech (2010) 2(5):S35-S37
ISSN: 0974-2328

Experimental methods for registration of electrical activity of the heart on cell and whole organ level

Ivo Provaznik*, Jana Kolarová, Jiri Sekora, Oto Janousek, Marina Ronzhina, Katerina Fialova, Marie Novakova

Received: 25 October 2010 / Received in revised form: 13 August 2011, Accepted: 25 August 2011, Published: 25 October 2011
© Sevas Educational Society 2011

Abstract

Modern recording systems for registration of electrical activity of the heart are based on application of voltage sensitive dye (VSD) into the examined tissue. Excited VSD's undergo changes in their electronic structure, and consequently their fluorescence spectra, in response to changes in the surrounding electric field. Fast VSDs are consistently sensitive probes and can be used in electrophysiological studies. We have proved that optical recording employing di–4–ANEPPS dye utilized in our laboratory is suitable registration method of electrical activity of isolated hearts perfused according to Langendorff. The heart is exposed to VSD diluted in Krebs–Henseleit solution to the concentration of $2\mu M$.

Keywords: heart, electrical activity, electro–optical recording systems

Introduction

Recently, electro-optical recording systems are often used to record electrical activity of isolated cardiomyocytes and the isolated heart perfused according to Langendorff (Bachtel et al. 2011). The systems are usually based on application of voltage-sensitive dyes into examined tissues. Generally, the VSD should exhibit large fluorescence and/or absorption changes that vary with changes of the membrane potential (Salama 2001). Further, its optical responses must be specifically related to the changes of membrane potential and not to ion concentration, transmembrane currents, or membrane conductance. Staining of the heart with the dye should

Ivo Provaznik*, Jana Kolarova, Jiri Sekora, Oto Janousek, Marina Ronzhina

Department of Biomedical Engineering, Faculty of Electrical Engineering and Communication, Brno University of Technology

Katerina Fialova, Marie Novakova

Department of Physiology, Medical Faculty, Masaryk University Brno

*Tel: +420 541 149 562, Fax: +420 541 149 542
E-mail: provaznik@feec.vutbr.cz

not result in pharmacological or toxic effects to the preparation. The response time of the dye has to lay in microsecond or submicrosecond range. For reliable and repeatable recording, the dye should be optically stable in its local environment and not prone to photobleaching upon an exposure to intense light (Novakova et al. 2008). Such systems can be extended with traditional touchless electrodes to record electrograms simultaneously with optical signals as recordings of electrical activity of individual myocytes.

Materials and Methods

The voltage sensitive dye bound to the cell membrane responds to the membrane potential change. A dye suitable for the measurement of electrical activity in cardiac preparations must respond to the potential changes with the time constant below 1/100 s. Dyes from ANEPPS group (amino–naphthyl–ethenyl–pyridinium) are the most constantly used in cardiac preparations (Nygren et al. 2003). Styryl-derived ANEPPS fast voltage-sensitive dyes were widely used for this purpose. These electrochromic dyes utilize CT mechanism from the electron-donating N,N-dialkylamino group to the electron-deficient pyridinium core in ground and excited states. As a result, the electrochromic dye responds to the membrane potential change by a spectral shift as shown in Fig. 2 (Kolarova et al. 2010). In an ideal case, the emission spectrum shifts bathochromically and does not change its shape or amplitude. The fluorescence intensity decreases with rising membrane potential. The combination of observed bathochromic shift and decay of the fluorescence intensity upon increasing membrane potential is used in so-called fluorescent emission ratio measurements.

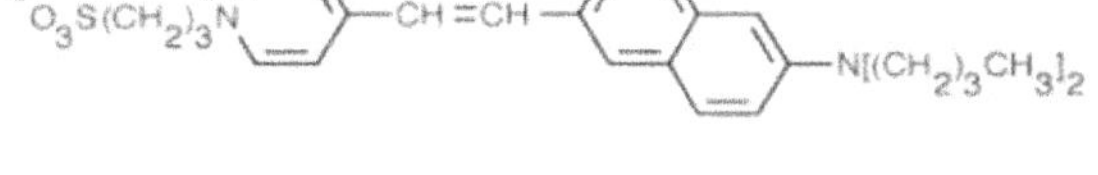

Figure 1: Aminonaphthylethenylpyridinium dye di-4-ANEPPS structure

Di–4–ANEPPS is utilized in our laboratory for recording of monophasic action potentials (MAPs) by optical method in various animal models. Most often employed experimental model is isolated heart perfused according to Langendorff. The heart is exposed to

VSD diluted in Krebs–Henseleit solution to the concentration of 2μM (Novakova et al. 2005).

The described system employs a flexible bifurcated fiber cable with illumination fibers positioned in a circle and a detection fiber positioned in the center of the cable (Provaznik et al. 2004). The optical probe is attached to the preparation to suppress motion artifacts without a need of focusing. The input end of the cable with six illumination fibers is connected to a light source. The output (detection) fiber is connected to a light detector that senses the beam of emitted light.

A halogen light source was chosen as a source of an excitation light. This cold light source with high intensity light output is designed for fiber optic applications with stable light output. The light source contains a built-in IR filter, which prevents a preparation from heating, and a band-pass filter (560 nm +/- 30 nm), which selects light at excitation maximum of the used dye.

The changes in dynamics of transmembrane potential result in amplitude modulation of the emitted light. This is detected by a photodiode detector with a high-pass (>610 nm) filter. The output signal of the photodiode detector is preamplified so that the two stage amplifier adjusts the signal to input range of data acquisition card (±1 V). The electrical circuits include also an analogue anti-aliasing filter (low-pass filter fc=2 kHz) and a high-pass filter (fc=0.05 Hz) to suppress DC offset.

Orthogonal ECG signals are recorded from six silver-silver electrodes positioned on the inner surface of the bath. All signals from the light detector and electrodes are simultaneously digitized by 12-bit AD converters at 4000 samples/sec rate. The digital signals are stored on a hard disk for further off-line processing (noise suppression, peak detection, visualization and analysis).

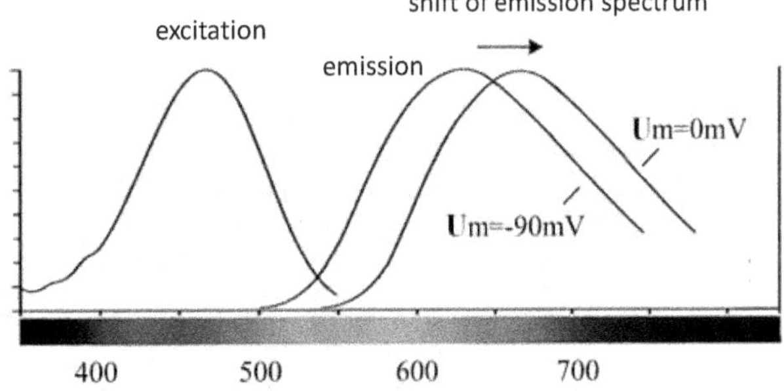

Figure 2: Optical characteristics of voltage sensitive dye di–4–ANEPPS

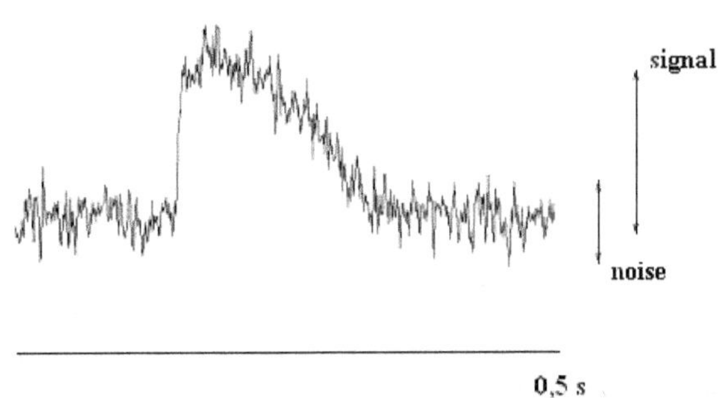

Figure 3: Raw monophasic action potential recorded by optical method from the spontaneously beating heart.

Discussion and conclusion

The possibility of recording of the transmembrane potential of excitable cells by optical methods was first suggested in 1968. The first cardiac application was then reported in 1981. Since then the method has been improved and numerous VSDs have been tested (Dillon et al. 1998). Several problems had to be solved before these new dyes were accepted for everyday laboratory practice (Svrcek et al. 2009). One of the most important tasks was to minimize side effects of the dye on the preparation in the absence and presence of light (Salama 2001; Salma et al. 2005). Most prominent pharmacological effect of VSDs on cardiac tissue is so-called photodynamic or phototoxic damage (Zochowski et al. 2000).

We have proved that optical recording employing VSDs is suitable registration method of electrical activity of isolated cardiomyocytes and isolated hearts. Certain electrophysiological changes are present in the hearts due to presence of dyes. Most prominent pharmacological effect of VSDs on cardiac tissue is photodynamic or phototoxic damage. The exact mechanism of these effects remains unexplained. Formation of free radicals or direct interaction with the voltage–gated calcium and/or potassium channels may result in altered conductivity and the time–dependent gating. However, these changes are mostly reversible and thus myocardium can be stained with di–4–ANEPPS and be considered a reliable model for electrophysiological studies.

Acknowledgement

This work was supported from: GAČR 102/07/1473, GAČR 102/09/H083, and MSM0021630513.

References

Bachtel AD, Gray RA, et al. (2011) A Novel Approach to Dual Excitation Ratiometric Optical Mapping of Cardiac Action Potentials With Di-4-ANEPPS Using Pulsed LED Excitation. IEEE Transactions on Biomedical Engineering, 58(7):2120-2126

Bardonova J, Provaznik I, Novakova M, et al. (2007) Statistical Analysis in Complex-Valued Wavelet Analysis of Voltage-Sensitive Dye Mapping. In Computers in Cardiology 34:101-104

Dillon S M, Kerner T E, Hoffman J, et al. (1998) A system for in-vivo cardiac optical mapping. IEEE Engineering in Medicine and Biology 95-108

Kolarova J, Fialova K, Janousek O, et al. (2010) Experimental methods for simultaneous measurement of action potentials and electrograms in isolated heart. Physiol Res, 59 (Suppl. 1):71-80

Novakova M, Bardonova J, Provaznik I, et al. (2008) Effects of voltage sensitive dye di–4–ANEPPS on guinea pig and rabbit myocardium. Gen Phys Biophys 27:45–54

Novakova M, Blaha M, Bardonova J, et al (2005) Comparison of the tissue response during the loading with voltage-sensitive dye in two animal models. In Computers in Cardiology 32, 535-538

Nygren A, Kondo C, Clark RB, et al. (2003) Voltage–sensitive dye mapping in Langendorff–perfused rat hearts. Am J Physiol Heart Circ Physiol 284:892–902

Provaznik I, Bardonova J, Novakova M, et al (2004) Analysis of optical recording stability using wavelet entropy of action potentials. In Proceedings of the 26th Annual International Conference of the IEEE Engineering in Medicine and Biology Society 26:377-379

Svrcek M, Rutherford S, Chen A, et al. (2009) Characteristics of Motion Artifacts in Cardiac Optical Mapping Studies. In Proceedings of the IEEE EMBS Conference 3240-3243

Salama G (2001) Optical mapping: background and historical perspective. In: Rosenbaum DS and Jalife J. Optical Mapping of Cardiac Excitation and Arrhythmias. Futura Publishing Company, Inc. Chapter 1:9-31

Salama G, Choi B R, Azour G, et al. (2005) Properties of new, long-wavelength, voltage-sensitive dyes in the heart. J Membr Biol 208:125-140

Zochowski M, Wachowiak M, Falk C X, Cohen L B, Lam Y W, Antic S, Zecevic D (2000) Imaging membrane potential with voltage-sensitive dyes. Biol Bull 198:1-21

J Biochem Tech (2010) 2(5):S38-S39
ISSN: 0974-2328

Thermodynamics of potential antioxidant action of Δ^5-sterols

Jozef Lengyel, Ján Rimarčík, Lenka Rottmannová, Erik Klein*

Received: 25 October 2010 / Received in revised form: 13 August 2011, Accepted: 25 August 2011, Published: 25 October 2011
© Sevas Educational Society 2011

Abstract

In this work, we have calculated bond dissociation enthalpies of selected C–H and O–H bonds in sterols using two semi–empirical quantum chemistry methods AM1 and PM3.

Keywords: sterols, BDE, hydrogen atom transfer, AM1, PM3

Introduction

Phytosterols are triterpene compounds. They are structural components of plant membranes and stabilize phospholipid bilayers in plant cell membranes. In last 10 years, they received considerable attention because of their cholesterol–lowering properties, anticancer and antioxidant activity (Moreau et al. 2002; White and Armstrong 1986). In this work, we have studied following phytosterol: Δ^5–sitosterol, Δ^5–campesterol, Δ^5–avenasterol, Δ^5–brassicasterol, Δ^5 –stigmasterol (Fig 1). We have studied also cholesterol present in animal cell membranes. Published works indicate that major targets of O_2 attack during sterol oxidation are C7–H and O–H bonds (Smith, 1996).

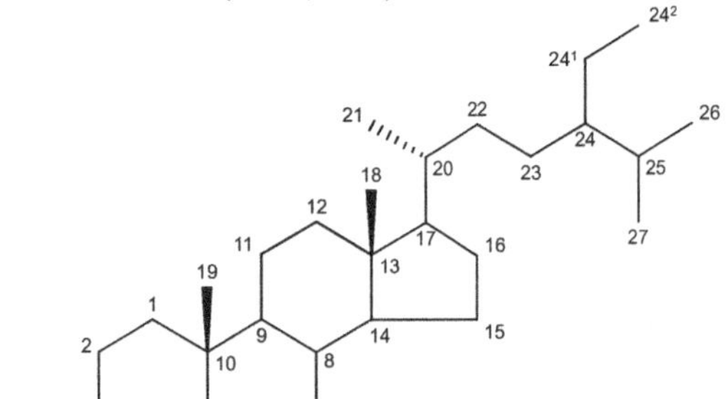

Figure 1: Atom numbering in sterols

Jozef Lengyel, Ján Rimarčík, Lenka Rottmannová, Erik Klein*

Institute of Physical Chemistry and Chemical Physics, Slovak University of Technology, Radlinského 9, SK–812 37 Bratislava, Slovak Republic

*Tel: +421 593 25 535, Fax: +421 52926032
E-mail: erik.klein@stuba.sk

Materials and methods

Geometry optimization of parent molecules and corresponding radicals was performed using PM3 and AM1 semiempirical quantum chemistry methods using Hyperchem program package (Hyperchem, USA). These are specially designed to obtain enthalpies of formation of chemical systems. Dewar et al. (1985) AM1 method (Austin Model 1) represents a standard tool for both theoretical and experimental organic chemists. Later, Stewart (1989) proposed a mathematical re-parameterization called PM3 (Parameterized Model 3). Bond dissociation enthalpies (BDE) were calculated using the formation enthalpies of the molecule, M, radical, R, and hydrogen atom, H

$$BDE = \Delta_f H(R) + \Delta_f H(H) - \Delta_f H(M)$$

Results and Discussion

Table 1 summarizes obtained results for C7–H and O–H bonds homolytic dissociation. These show that C7–H bond cleavage requires less energy than O–H bond. C7 carbon is in α–position to C5=C6 double bond. AM1 method predicts identical BDEs for all sterols. PM3 predicts BDE(C7–H) and BDE(O–H) values in 23 kJ mol^{-1} and 9 kJ mol^{-1} range, respectively. Since all sterols have identical structure of the rings, very close or identical BDEs can be expected. The results indicate that different tails in the studied molecules, attached at C17, do not affect BDEs. BDE(C7–H) are similar to O–H BDEs found for four tocopherols (Klein et al. 2007) that represent most powerful natural antioxidants.

Table 1: Bond dissociation enthalpies of studied sterols

Δ^5-sterol	BDE(C7–H) kJ mol^{-1}		BDE(O–H) kJ mol^{-1}	
	PM3	AM1	PM3	AM1
Sitosterol	330	336	421	466
Campesterol	330	336	422	466
Avenasterol	322	336	430	466
Brassicasterol	323	336	430	466
Stigmasterol	345	336	42	466
Cholesterol	330	336	421	466

Conclusion

In accordance with experimental works (Lampi et al. 2002; Smith 1996), we can conclude that most important site of oxidation of studied sterols is hydrogen at C7 atom.

Acknowledgement

The work has been supported by Scientific Grant Agency (VEGA Project 1/0137/09).

References

Dewar MJS, Zoebisch EG, Healy EF, Stewart JJP (1985) Development and use of quantum mechanical molecular models. 76. AM1: a new general purpose quantum mechanical molecular model. Journal of American Chemical Society 107(13):3902-3909

Klein E, Lukeš V, Il čin M (2007) DFT/B3LYP study of tocopherols and chromans antioxidant action energetics. Chemical Physics 336(1):51-57

Lampi AM, Juntunen L, Toivo J, Piironen V (2002) Determination of thermo-oxidation products of plant sterols. Journal of Chromatography B-Analytical Technologies in the Biomedical and Life Sciences 777(1-2):83-92

Moreau RA, Whitaker BD, Hicks KB (2002) Phytosterols, phytostanols, and their conjugates in foods: structural diversity, quantitative analysis, and health-promoting uses. Progress in Lipid Research 41(6):457-500

Smith LL (1996) Review of progress in sterol oxidations: 1987-1995. Lipids 31(5):453-487

Stewart JJP (1989) Optimization of parameters for semiempirical methods II. Applications. Journal of Computational Chemistry 10(2):221-264

White PJ, Armstrong LS (1986) Effect of selected oat sterols on the deterioration of heated soybean oil. Journal of the American Oil Chemists Society 63(4):525-529

J Biochem Tech (2010) 2(5):S40-S41
ISSN: 0974-2328

The role of various transport layers type for construction of organic solar cells

Martin Sedina, Patricie Heinrichova, Martin Weiter, Martin Vala, Ota Salyk*

Received: 25 October 2010 / Received in revised form: 13 August 2011, Accepted: 25 August 2011, Published: 25 October 2011
© Sevas Educational Society 2011

Abstract

The contribution discusses the role of various types of charge transport layers used in two different architectures of organic solar cells. Parameters of devices based on those architectures are compared.

Keywords: charge transport layers, organic solar cells

Introduction

The organic solar cells are usually based on blends of conjugated polymers, which mainly represent electron donor, and of high or low molecular–weight organic materials operating as electron acceptor (Al–Ibrahim and Sensfuss 2005). Layers of these materials can be prepared from solutions. That provides low–cost of fabrication of photovoltaic devices made from organic materials. Recent reported conversion efficiency of organic solar cell is about 8 %. Therefore, it is necessary to improve the performance of the devices. Useful way seems to be an innovation in manufacturing methods, which mainly depends on the architecture of the cells. In this contribution, we report solar cells based on inverse architecture and compare their performance with the "normal" organic solar cells.

In the inverse structure, the photogenerated charge carriers are collected by opposite electrodes as for the traditional type and it can significantly affect some of the basic parameters of solar cells prepared from polymeric or low molecular–weight materials.

Materials and methods

We prepared samples with traditional structure ITO/PEDOT: PPS/MDMO–PPV:PCBM/Al, and ITO/PEDOT:PPS/Pc/PCBM/Al

Martin Sedina, Patricie Heinrichova, Martin Weiter, Martin Vala, Ota Salyk*

Brno University of Technology, Faculty of Chemistry, Centre fo r Materials Research CZ.1.05/2.1.00/01.0012, Purkynova 464/118, 612 00 Brno, Czech Republic

*Tel: + 420 541 149 397, Fax: + 541 211 697
E-mail: xcsedina@fch.vutbr.cz

and samples in inverse configuration (ITO/TiO$_2$/MDMO–PPV:PCBM/Au and ITO/TiO$_2$/PCBM/Pc/Au). In the inverse structure, the photogenerated charge carriers are collected by opposite electrodes as for the traditional type. The basic parameters characterising solar cells can be obtained from current–voltage characteristics during illumination. These I–V characteristic were measured by Keithley 6517 Electrometer, that could record currents from 2pA to 200mA and voltage from 200 mV to 200V. A 450 W xenon lamp was used as a source of illumination.

ITO – Indium tin oxide,
PEDOT:PSS–Poly (3,4–ethylenedioxylthiophene) poly(styrenesulfonate),
TiO$_2$ – Titanium dioxide,
MDMO–PPV–Poly (2–methoxy–5–(3'–7'–dimethyloxy)–1,4–phenylenvinylene,
Pc – Phtalocyanine.

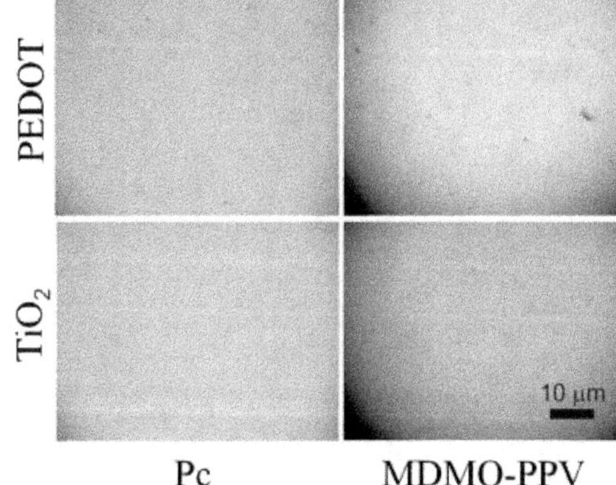

Figure 1: Microphotographs of functional layers of both materials prepared on different transport layers, the PEDOT and TiO$_2$.

Results and Discussion

The microphotographs on Fig 1 show that the morphology of the prepared functional layers of Pc and MDMO-PPV on both transport materials is not very different from each other and that the actual choice of transport layer doesn't significantly affected the internal organization of subsequent functional layers. All these functional layers have approximately the same thickness around 100 nm.

Electrical measurements were performed on the prepared organic solar cells and the I–V characteristics were obtained. Regions, where the current is negative and the voltage is positive were selected from these measurements. In this region, the solar cells supplied electricity into the external circuit.

These 4th quadrant of I–V characteristics of normal and inverse organic solar cells under illumination with xenon lamp (intensity of illumination $P_0 = 1$ mW/cm^2) are shown in Fig. 2a and Fig. 2b. From the graphs, it can be seen, that in this part of the I–V characteristics a transition from convex to concave function was achieved due to the change of transport layer for both types of functional materials. This is reflected by the fact that in both cases the fill factor (FF) increased in comparison with traditional organic solar cells. This improvement may be associated with a better interface between the electrode and the layer responsible for maintaining the charge (Petritsch 2000).

2a)

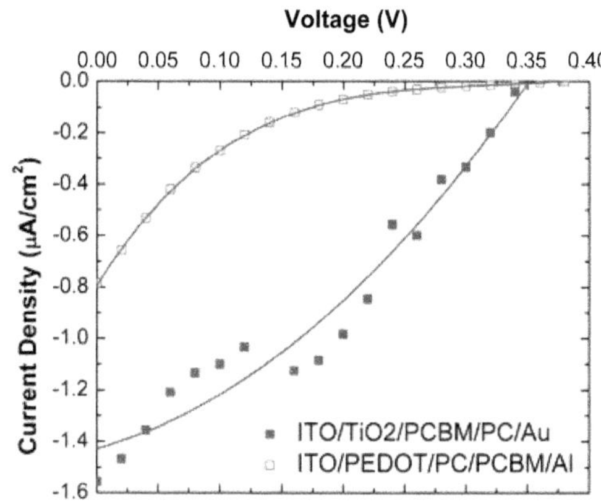

2b)

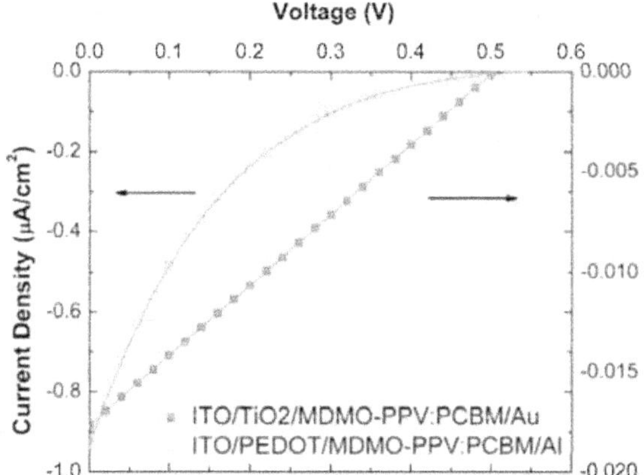

Figure 2: 4th quadrant of I–V characteristics of cells with functional layer of a) Pc and b) MDMO–PPV

Table 1: Parameters summarising the performance of the solar cells

Cell	V_{OC} (mV)	I_{SC} (μA/cm²)	FF (%)	n (%)
ITO/PEDOT/Pc/PCBM/Al	381	0.791	9.0	0.152
ITO/TiO2/PCBM/Pc/Au	342	1.498	35.9	1.027
ITO/PEDOT/MDMO–PPV+PCBM/Al	537	0.932	10.3	0.286
ITO/TiO2/MDMO–PPV+PCBM/Au	502	0.018	25.1	0.012

Table 1 summarizes the most important parameters of both types of solar cells. In the case of phthalocyanine in inverse structure an increase of short-circuit current density, a slight reduction in open-circuit voltage and mostly, a significant growth of fill factor were observed. This caused an increase of the efficiency of energy conversion up to $\eta = 1.027\%$. For the solar cells based on the polymer MDMO-PPV, an inverse structure appears to be less useful, because there was a significant decrease in short-circuit current density, which had a major influence on the decrease of the energy conversion efficiency ($\eta = 0.286$ %) for the standard configuration and $\eta = 0.012$ % for the inverse one.

Conclusion

We managed to prepare organic solar cells in standard and inverse structure (thanks to the use of different transport materials). As a functional layer we used both, low-molecular weight substances (phthalocyanine) and high-molecular weight polymer (MDMO-PPV).

The electrical measurements show that both of the inverse devices have higher fill factor (FF) than traditional solar cells. The inverse device based on phthalocyanine has significantly improved efficiency of energy conversion (up to $\eta = 1.027\%$). This improvement may be associated with a better interface between the electrode and the layer responsible for extraction of the charge.

Acknowledgement

The work has been supported by Grant Agency of the Czech Republic via project No. P205/10/2280.

References

Al–Ibrahim M, Sensfuss S (2005) Solar Energy Materials & Solar Cells, 85:277–283

Petritsch K (2000) Organic Solar Cell, Cambridge and Graz. 142

J Biochem Tech (2010) 2(5):S42-S43
ISSN: 0974-2328

Electrochemical determination of the effect of deoxynivalenol on rats

Pavlina Sobrova, Anna Vasatkova, Vojt ech Adam, Sona Krizkova, Miroslava Beklova, Ladislav Zeman, Rene Kizek

Received: 25 October 2010 / Received in revised form: 13 August 2011, Accepted: 25 August 2011, Published: 25 October 2011
© Sevas Educational Society 2011

Abstract

This study is focused on examining the impact mouldy wheat contaminated with deoxynivalenol (DON) on the health status of rats and their ability to resist oxidative stress. Among the factors determining the defences include the activities of liver detoxification enzymes and content of blood peptides and proteins (glutathione and metallothionein).

Keywords: Deoxynivalenol, rats, metallothionein, *Fusarium*

Introduction

Mycotoxins are secondary metabolites produced by micro-fungi that are capable of causing disease and death in humans and other animals. While aflatoxins are undoubtedly the most important mycotoxins, *Fusarium* toxins have recently been given more attention (Pestka et al. 1990). The biggest risk is associated with the occurrence of a deoxynivalenol (DON). After ingestion of mycotoxin there are activated numerous mechanisms in a body to

Pavlina Sobrova, Vojtech Adam, Sona Krizkova, Rene Kizek*

Department of Chemistry and Biochemistry, Faculty of Agronomy, Mendel University in Brno, Zemedelska 1, CZ–613 00 Brno, Czech Republic

*Tel: +420 545 133 350, Fax: +420 545 212 044
E-mail: kizek@sci.muni.cz

Anna Vasatkova, Ladislav Zeman

Department of Animal Nutrition and Forage Production, Faculty of Agronomy, Mendel University in Brno, Zemedelska 1, CZ–613 00 Brno, Czech Republic

Miroslava Beklova

Department of Veterinary Ecology and Environmental Protection, Faculty of Veterinary Hygiene and Ecology, University of Veterinary and Pharmaceutical Sciences, Palackeho 1–3, CZ–612 42 Brno, Czech Republic

reduce their toxicity and to induce their subsequent excretion (Jajic et al. 2008). Very important substances providing their detoxifications are thiol compounds contained in metallothionein and glutathione.

Electrochemical methods for metallothionein determination are very suitable methods for protein determination due to the very low limit of detection.

Materials and methods

This work was aimed at investigation of influence of DON on rats treated with deoxynivalenol for 28 days. The effect of DON was revealed by voltammetric and spectrometric serving for determination of reduced and oxidized glutathione, metallothionein and activities of certain enzymes connected with detoxification of xenobiotics in serum and tissues of rats.

Selected male Wistar albino laboratory rats of 28 days of age were used in our experiments (Fig 2). Experimental animals were kept in a vivarium with controlled air temperature ($23 \pm 1°C$) and photo-period (12 hours day:12 hours night with maximal intensity 10800 LUX). Tempered feed mixtures and drinking water were accessible ad libitum. Animals were weighed once a week and at the same time weight gain, feed intake, conversion and health state were monitored. Feed conversion was calculated according to the following equations: (feed intake)/(weight gain). The experimental animals were divided into eight groups (eight male rats per group). We used feed mixtures of natural barley and barley contaminated by mycotoxins both with different content of vitamins, organic and inorganic zinc (II) form (Table 1). Natural barley contained $40 \pm 5 \mu g/kg$ of DON and naturally contaminated barley $3500 \pm 5 \mu g/kg$ of DON. The rats were administrating these mixtures for 28 days. In the end of the experiment, the animals were put to death and tissues and blood were sampled.

Preparation of Biological Samples

The rats' tissues (liver, kidney, spl een, heart, brain, eye, gonads and femoral muscle) and blood were used for the analysis. The animal tissues were mixed with extraction buffer (100mM potassium phosphate, pH 6.8) and subsequently homogenized using polo automatic homogenisator (Shutt homogen plus, Germany).

Results and Discussion

We found that administration of deoxynivalenol caused a response in the form of increased synthesis of metallothionein. Another interesting area was the effect of increasing doses of zinc, which was added both to the mouldy and to "control" diet. Synergy effect of mycotoxins is very interesting and very little explored area. The possible link between zinc intake and its influence on the detoxification of mycotoxins via enhanced synthesis of metallothionein, whose transcription is activated by zinc ions presence, seems to be very interesting. Levels MT in different tissues varied. The highest concentration of MT was found in tissues closely connected with the detoxification of xenobiotics, such as kidney (6.69 ± 0.05 μg/ml) and liver (6.06 ± 0.05 μg/ml). Half concentration was detected in heart, brain, testes and muscle (approximately 3.0 μg/ml).

Furthermore, the MT levels differed in individual organs and the difference between groups feeding by natural and mouldy barley was observed (p<0.05) except spleen and brain where the difference was not significant. It was observed the similar trend of MT level in single groups. In both groups the highest level had specimens, which feedstuff was enriched with organic chelate zinc (II) form. Significant (p<0.05) higher was MT level at feedstuff with organic chelate in liver, kidney and muscle. DON-containing group was slightly increased against the non-DON group. Evident was interaction (p<0.01) at gonads. DON-containing group showed after vitamins, organic and inorganic zinc supplementation lower MT level with comparison non-DON group. Inorganic form also increased the MT expression but the results did not reach the values of organic zinc. Feed supplemented by vitamins did not indicate the effect after consumption. Moreover, organic and inorganic zinc (II) and vitamins supplementation have no significant effect on MTs level in brain tissues of rats. To verify the electrochemical detection the Sodium dodecyl sulphate polyacrylamide gel electrophoresis was used. The results obtained from electrochemical detection were similar as the results from gel electrophoresis.

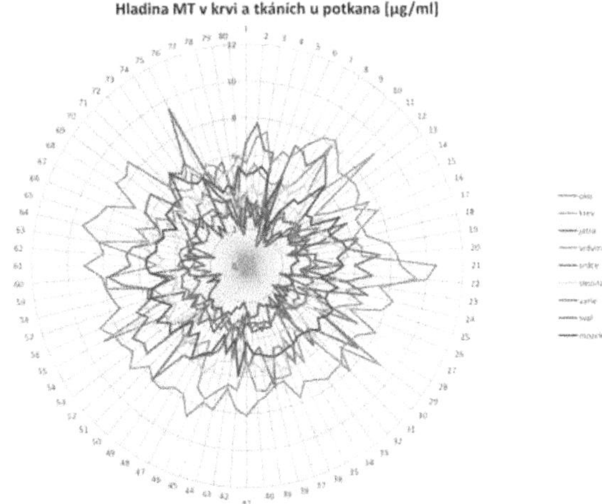

Figure 1: MT levels in single tissues.

Conclusion

Studying the impact of toxic substances in organisms is still very current and provides very interesting results. Toxic substances that are part of our diet, are the more objective of such studies because of the possible benefits. We are able not only to define the level of toxicity, but also gain an idea of the biochemical effect of the studied substance, which can bring new opportunities in its detoxification.

Acknowledgement

This work was supported by grant MSMT 6215712402 and GA AV IAA401990701.

References

Jajic I, Juric V, Glamočic D, Abramovic B (2008) Occurrence of Deoxynivalenol in Maize and. Wheat in Serbia Int J Mol Sci 9(11):2114–2126

Pestka JJ, Bondy GS (1990) Alteration of immune function following dietary mycotoxin exposure. Can J Physiol Pharmacol 68(7):1009–1016

J Biochem Tech (2010) 2(5):S44-S45
ISSN: 0974-2328

Tayloring of molecular materials for organic electronics

Martin Vala*, Martin Weter, Patricie Heinrichova, Martin Sedina, Imad Ouzzane, Petra Moziskova

Received: 25 October 2010 / Received in revised form: 13 August 2011, Accepted: 25 August 2011, Published: 25 October 2011
© Sevas Educational Society 2011

Abstract

An example of targeted modification of chemical structure in order to extent utilisation of diketo–pyrrolo–pyrroles in organic electronics is presented. The introduction of solubilising groups opens solution based techniques for device preparation and therefore reduces the manufacturing cost. The influence of the substitution on the optical activity is discussed.

Keywords: Organic electronics, diketo–pyrrolo–pyrroles

Introduction

Diketo-pyrrolo-pyrroles (DPPs) comprise one of the promissing classes of materials suitable for organic electronics. Nowadays, the DPPs are mainly used as high performance industrially important pigments. This requires excelent photostability, high absorption coeficients but also low solubility. However, the ability to solubilize these materials would open the possibility to use solution–based techniques (spin–coating, drop–casting, inject printing, etc.) to prepare devices from DPPs. One of the reason for the insolubility is the existence of hydrogen bonds between the –NH group and oxygen. Since the basic DPP core is perfectly planar, a π-π electrons overlap occurs in solid state and also contributes to their insolubility. These interactions can be so strong, that cause colour change between solid and dissolved form and influence also other properties, e.g. fluorescence and Stokes shift (Song et al. 2007). It is therefore clear, that to modify the solubility one has to introduce the *N*–substituion and/or break the molecule planarity (Potrawa and Langhals 1987). In this contribution, we will discuss the influence of such substitution on the optical properties of the DPPs with respect to the utilization in organic electronics and possibly in bielectronics in future.

Martin Vala*, Martin Weiter, Patricie Heinrichova, Martin Sedina, Imad Ouzzane, Petra Moziskova

Brno University of Technology, Faculty of Chemistry, Purky ňova 464/118, Brno 61200, Czech Republic

*Tel: +420 541 149 411, Fax: +420 541 211 697
E-mail: vala@fch.vutbr.cz

Materials and Methods

All of the studied derivatives (Fig 1) were synthesised by Research institute of organic synthesis (RIOS, a.s.). The samples were dissolved in dimethylsulfoxide (Aldrich).

Results and Discussion

Substitution of an alkyl group on the nitrogen on the DPP core decreases molar absorption coefficient (hypochromic shift) and simultaneously the longer wavelength maximum is shifted towards higher energy region (hypsochromic shift). Furhermore, the vibration structure is less pronounced. As was pointed out in our previous paper reporting different structures (Vala et al. 2008), this is caused by torsion between pyrrolinone central part and phenyl adjacent to the alkyl group and consequently, is caused by loss of molecule planarity which is in turn responsible for loss of effective conjugation. Since the addition of second alkyl rotates also the second phenyl group, this effect is even more pronounced. The loss of vibration structure can be attributed to the increased dipole moment interacting with polar DMSO solvent. The dipole–dipole interaction of bi–substituted derivatives with the completely non-planar structure is the most pronounced. No dependency on the length of the alkyl used was found.

Figure 1: The studied derivaties of diphenyl–diketopyrrolo– pyrrole

Conclusions

It was found, that the *N*–alkylation only does not significantly influence the fluorescence quantum yield. On the other hand the Stokes shift is gradually increased going from the monoalkylated to dialkylated derivatives. The observed spectra are characteristic by graduate loss of mirror symmetry of absorption–fluorescence and vibronic structure. The phenyl torsion due to the *N*–alkylation is the main mechanism for this behaviour in polar DMSO.

Acknowledgement

The work was supported by Grant Agency of the Czech Republic via project No. P205/10/2280.

References

Potrawa T, Langhals H (1987) Fluorescent dyes with large stokes shifts - soluble dihydropyrrolopyrrolediones. Chemische Berichte-Recueil 120(7):1075-1078

Song B, Wei H, Wang ZQ, Zhang X, Smet M, et al. (2007) Supramolecular nanoribers by self-organization of bola-amphiphiles through a combination of hydrogen bonding and pi-pi stacking interactions. Advanced Materials 19(3):416

Vala M, Weiter M, Vynuchal J, Toman P, Lunak S (2008) Comparative Studies of Diphenyl-Diketo-Pyrrolopyrrole Derivatives for Electroluminescence Applications. Journal of Fluorescence 18(6):1181-1186

J Biochem Tech (2010) 2(5):S46-S47
ISSN: 0974-2328

Acute toxicity of lanthanum to fish *Danio rerio* and *Poecilia reticulata*

Stanislava Mácová, Petr Babula, Lucie Plhalová, Krystýna Žáčková, René Kizek*, Zdeňka Svobodová

Received: 25 October 2010 / Received in revised form: 13 August 2011, Accepted: 25 August 2011, Published: 25 October 2011
© Sevas Educational Society 2011

Abstract

Lanthanides are widely used in industry as well as medicine, where serve as contrast agents in imaging methods and techniques. Due to these facts, lanthanides can become significant pollutants of living environment. In our work, we were focused in investigation of toxicity of lanthanum on juvenile stages of two fish species- *Danio rerio* and *Poecilia reticulata* and embryonic stages of *D. rerio*. The 96hLC50 values were 156.33 ± 5.59 mg.l^{-1} for juvenile *D. rerio* and 128.38 ± 5.29 mg.l^{-1} for juvenile *P. reticulata*. The 144hLC50 for embryonic stages of *D. rerio* was 152.98 ± 8.06 mg.l^{-1}. Results of toxicity tests indicate possible toxicity of lanthanum in the case of presence in aquatic environment.

Keywords: Zebrafish, guppy, embryonic stages, juvenile stages

Introduction

Lanthanides include group of elements from lanthanum to lutetium, which are widely used in industry. Lanthanides (III) ions have many biological properties, which are based especially on similarity with calcium ions. These properties include blocking of calcium channels, which results in inhibition of contraction of skeletal,

---

Stanislava Macova, Lucie Plhalova, Zdenka Svobodova

Department of Veterinary Public Health and Toxicology, Universit y of Veterinary and Pharmaceutical Sciences, Palackeho 1–3, CZ–612 42 Brno, Czech Republic

Petr Babula, Krystyna Zackova

Department of Natural Drugs, University of Veterinary and Pharmaceutical Sciences, Palackeho 1–3, CZ–612 42 Brno, Czech Republic

Rene Kizek*

Department of Chemistry and Biochemistry, Mendel University in Brno, Zemedelska 1, CZ–613 00 Brno, Czech Republic

*Tel: +420 545 133 350, Fax: +420 545 212 044
E-mail: kizek@sci.muni.cz

smooth and cardiac muscles, replacing of calcium(II) ions is structure of many proteins, which means that these proteins can lose their functions, or, contrariwise, in some cases their function can be activated or increased. Possibility of interactions with nucleic acids was also demonstrated (Kohoutkova et al. 2009). Some sources indicate that lanthanides are non–toxic after oral application because of inability to cross bell biomembranes, but these results are contradictory to results of other works (Fricker et al. 2006). Lanthanides are used as contrast agents, but some complexes find utilization in therapy of cancer, because of their significant cytostatics properties (Kynast et al. 2004; Heffeter et al. 2004). Due to these facts, it is evident that lanthanides represent potential source of living environment contamination. Our work was focused on testing of toxicity of lanthanum(III) ions on juvenile stages of *Danio rerio* and *Poecilia reticulata* and embryonic stages of *D. rerio*.

Materials and methods

Parameters of used tap water

The basic physical and chemical parameters of the dilution water used in toxicity tests on embryonic and juvenile stages were: ANC$_{4.5}$ (acid neutralization capacity) 3.6–3.7 mmol.l^{-1}; COD$_{Mn}$ (chemical oxygen demand) 1.4–1.9 mg.l^{-1}; total ammonia below the limit of determination (< 0.04 mg.l^{-1}); NO$_3^-$ 24.5–31.4 mg.l^{-1}; NO$_2^-$ below the limit of determination (< 0.02 mg.l^{-1}); Cl$^-$ 18.9–19.1 mg.l^{-1}; Σ Ca $\pm$ Mg 3.1 mmol.l^{-1}. Test solutions were made from a stock solution by using LaCl$_3$.

Acute toxicity tests

Acute toxicity tests on juvenile stages of zebrafish (*D. rerio*) and guppy (*P. reticulata*) (2–3 months old) were performed according to the OECD method No. 203 (Fish, acute toxicity test). Fish were acclimatized 72 hours before the tests in the tap dilution water under the standard conditions. Used concentrations of lanthanum were 75, 100, 125, 150, and 175 mg.l^{-1}. In each concentration and control were placed 10 fish which were randomly picked from the spare stock. Six series were made with *D. rerio* and 4 series with *P. reticulata*. The tests were made using a semi–static method with the solution replacement after 24 hours. During the tests, records of the water temperature, pH, the concentration of oxygen dissolved in test tanks and fish mortality were noted. Duration of each test was 96 hours. The temperature of the experimental bath was 23.0 ± 1.5 °C, pH was between 7–8, and dissolved oxygen concentration did not fall below 60%. No fish died in the control tanks during the experiments.

Embryo toxicity tests

Embryo toxicity tests were made according to the method OECD No. 212 (Fish, short–term toxicity test on embryo and sac–fry stages). The fertilized eggs of *D. rerio* were placed in Petri dish within 8 hours at the latest after fertilization. Five series, each with five concentrations of tested substance (100, 125, 150, 175, 200; 50, 100, 150, 200, 250 mg.l^{-1}) were used. Twenty embryos in Petri dish were tested at each concentration and in control. The semi–static method with the replacement of the tested solution after 24 hours was used. The tests were terminated after hatching of all individuals in the control Petri dish which was 144 h after the start of the test. Hatching and survival of embryos were recorded during the tests in 24 h intervals. Test bath temperatures were between 24.5 and 25.5°C. The mortality rate of the control embryos did not exceed 20%.

Results processing

The results of the toxicity tests (the number of dead individuals at particular test concentrations) were subjected to a probit analysis using an EKO–TOX 5.2 programme to determine the LC50 values of lanthanum. The statistical significance of the difference between LC50 values for the juvenile and the embryonic stages of *D. rerio* and the juvenile stages of *P. reticulata* was calculated using the non–parametric Mann–Whitney test and the Statistica v8.0 for Windows (StatSoft, USA).

Results and Discussion

We determined that lanthanum toxicity depends on used fish species; average LC50 after 96–hour exposition was 156.33 ± 5.59 mg.l^{-1} for *D. rerio* and 128.38 ± 5.29 mg.l^{-1} for *P. reticulata* . In the case of embryo toxicity tests on *D. rerio*, LC50 value 152.98 ± 8.06 mg.l^{-1}after 144–hour expositions was determined. The sensitivity of juvenile and embryonic stages of *D. rerio* was comparable. The highest sensitivity was found for juvenile stages of *P. reticulata* in comparison of juvenile stages of *D. rerio* (p<0.01) (Fig. 1).

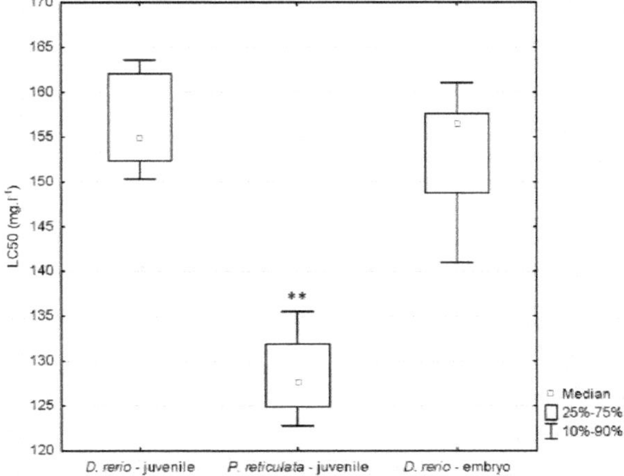

Figure 1: 96hLC50 values of juvenile stages of *D. rerio* and *P. reticulata* and 144hLC50 values of embryonic stages of *D. rerio*

Different results were presented by Zhang (2008) who stated 96hLC50 value 11.49 mg.l^{-1} for *Cyprinus carpio*. Chronic toxicity exposition (4 months) of lanthanides(III) ions of carp caused anomalies of erythrocytes (Zhang, 2008). Toxicity of lanthanides was demonstrated also for *Daphnia carinata* and *Artemia salina* (Barry and Meehan 2000; Cunqi 1998). Barry and Meehan (2000) assessed acute and chronic toxicity of lanthanum for *D. carinata* in three different media. The 48h EC50 to *Daphnia* was 43 µg.l^{-1} in soft tap water, 1180 µg.l^{-1} in ASTM hard water and 49 µg.l^{-1} in the daphnid growth medium, based on diluted sea water. The chronic toxicity was measured in the daphnid growth medium and ASTM medium. There was found 100% mortality at concentration ≥80 µg.l^{-1} by after six days of the experiment in daphnid growth medium. In ASTM media was observed significant mortality at concentrations ≥39 µg.l^{-1}. Lower acute toxicity (66.47 mg.l^{-1}) of La(NO$_3$)$_3$·6H$_2$O was observed for *A. salina* after 96 hours exposure (Cunqi 1998). These results indicate possible toxicity of lanthanum ions in the case of presence in aquatic ecosystem.

Conclusion

In our study, lanthanum demonstrated toxicity on two fish species-*D. rerio* and *P. reticulata* and two developmental stages of *D. rerio*. There are some important questions, which must be solved in future; these questions include especially mechanism of toxic action of lanthanum ions including histochemical evaluation of fish tissues and mechanism of transport of lanthanum ions through biomembranes.

Acknowledgement

The work has been supported by grants: MSM Project No. 6215712402.

References

Barry M, Meehan BJ (2000) The acute and chronic toxicity of lanthanum to *Daphnia carinata*. Chemosphere 41(10):1669-1674

Cunqi H (1998) The actuate toxicitive effects of lanthanum and cerium on brine shrimp. Journal of Hebei University 18(1):68-79

Fricker SP (2006) The therapeutic application of lanthanides. Chemical Society Reviews 35(6):524–533

Heffeter P, Jakupec MA et al (2004) Toxicity and anticancer activity of the new lanthanum–centred compound KP772 (FFC24). Pharmacology 72(2):154-154

Kohoutkova V, Babula P, et al (2009) Study of DNA interactions with cerium (III), lanthanum (III) and gadolinium (III) ions by using of Raman spectroscopy. Febs Journal 27(6):95-95

Kynast B, Graf N, et al (2004) Novel tumor–inhibiting metal complexes: Preclinical efficacy and toxicity of the lanthanum(III) complex tris (1,10–phenantroline)–lanthanumltrithiocyanate FFC24. Journal of Clinical Oncology 22(14):3166-3166

Zhang GS (2008) Effects of lanthanum on the genetic toxicity and circulating blood corpuscle parameters in *Cyprinus carpio*. Journal of Anhui Agricultural Science 17

J Biochem Tech (2010) 2(5):S48-S49
ISSN: 0974-2328

Determination of heavy metals in fish scales

Hana Nováková, Markéta Holá, Jozef Kaiser, Jiří Kalvoda, Viktor Kanický*

Received: 25 October 2010 / Received in revised form: 13 August 2011, Accepted: 25 August 2011, Published: 25 October 2011
© Sevas Educational Society 2011

Abstract

The outcomes from measurements of amount of selected elements in the fish scales of common carp are presented. Concentrations in the scales were identified and differences between storage of heavy metals in exposed and covered part of scale were studied. The spatial distribution of elements on the fish scale's surface layer was measured by Laser Ablation–Inductively Coupled Plasma–Mass Spectrometry (LA–ICP–MS). The average amount of elements in the dissolved scales was quantified by ICP-MS. The fine structure of fish scales was visualized by phase–contrast Synchrotron radiation (SR) microradiography.

Keywords: fish scales, ICP-MS, LA-ICP-MS

Introduction

The information about an elemental distribution in fish scales can give insight to the degree of water pollution in the period of the fish life.

Hana Nováková, Markéta Holá, Viktor Kanický*

Department of Chemistry, Faculty of Science, Masaryk University, Kotlářská 2, 611 37 Brno, Czech Republic

*Tel: +420 549 494 774, Fax: +420 549 49 2443
E-mail: viktork@sci.muni.cz

Jozef Kaiser

Institute of Physical Engineering, Faculty of Mechanical Engineering, Brno University of Technology, Technická 2896/2, 616 69 Brno, Czech Republic

Jiří Kalvoda

Department of Geological Sciences, Masaryk University, Kotlářská 2, 611 37 Brno, Czech Republic

Fish scales have characteristics that can be found in other similar structures e.g. in bones and teeth. All these materials has as main components collagen type organic ingredients, hydroxyapatite ($Ca_{10}(PO_4)_6(OH)_2$) as inorganic components and water (Torres et al. 2008; Onozato et al. 1979). The lattice of hydroxyapatite contains a small amount of sodium, magnesium and carbonate ions. Each fish scale consists of two distinct layers: external, bony layer and inner fibrous layer (Ikoma et al. 2003).

Using LA–ICP–MS the distribution of matrix elements (Ca, Mg, P) and trace elements (Sr, Ba, Zn, Mn, Fe) was monitored. The choice of investigated chemical elements was motivated by the importance related to biological and environmental exposure.

Materials and methods

Measurements were performed at two fish scales from common carp (*Cyprinus carpio*) living in the Brno reservoir in South Moravia, Czech Republic. Before the analysis the samples had to be cleaned of debris, and a thin transparent skin, which secretes mucus, has to be removed. The impurities were relieved using ultrasound machine by submerging the scale in a 5% hydrogen peroxide solution for 5 minutes. All impurities were then removed under a microscope using a toothbrush. The process was repeated until the scale became completely clean.

The thickness of the scales and the individual layers were measured from the scale cross section by Backscattering Electron Microscopy (BSE). The thickness of the external layer changes from 50 to 100 μm, the collagen plate was 70 μm thick.

The samples were further analyzed on the LA–ICP–MS instrument, which is composed of: Nd: YAG laser ablation system UP 213 (New Wave Research, USA) emitting laser light at wavelength 213 nm and ICP–MS spectrometer Agilent 7500ce (Agilent, Japan) in a horizontal arrangement.

The optimized parameters for line scan across the sample were: the laser beam diameter – 40 μm, frequency 10 Hz, energy 8.1 J.cm^{-2} and the speed that the sample moved with – 40 μm.s^{-1}. In depth profiling measurement the following parameters were set ablation spot diameter 100 µm, laser frequency 4 Hz, energy 3.4 J.cm^{-2} for 90 s.

Results and Discussion

The following isotopes of elements, with expected uniform distribution across the fish scale were measured: ^{23}Na, ^{26}Mg, ^{31}P, ^{86}Sr and ^{42}Ca. The example of line scans of Ca and P is shown in Fig 1. On the base of measurements of heavy metals isotopes, we found that heavy metals are usually stored in more exposed parts of scales. This can be explained by the fact that these parts are more exposed to the water.

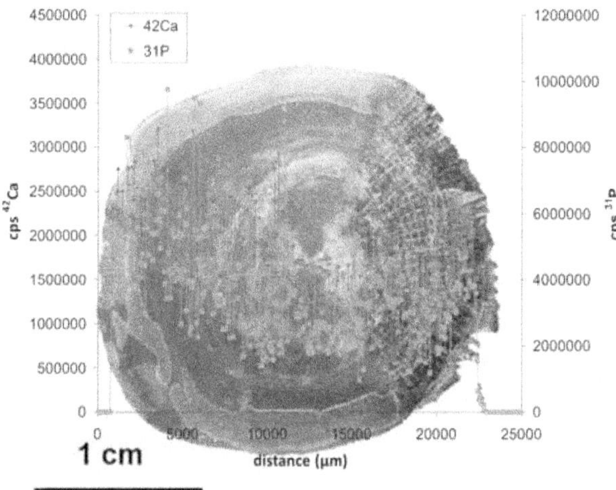

Figure 1: Line scans: intensity of ^{31}P and ^{42}Ca isotopes

Depth profiling, clearly demonstrated the differences on a signal obtained from the hydroxyapatite in collagen layers (Fig 2).

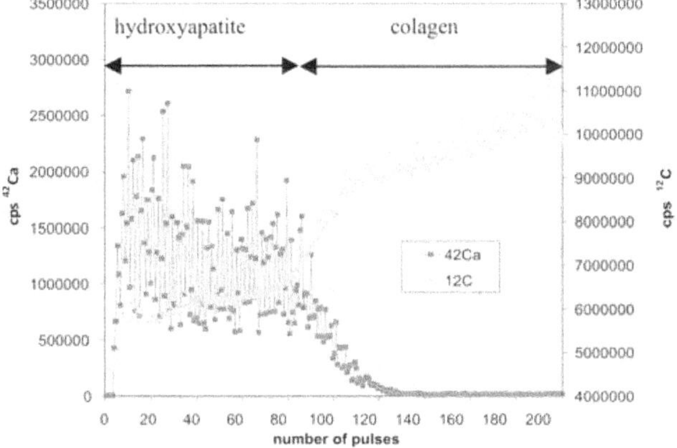

Figure 2: Depth profiling: intensity of ^{42}Ca and ^{12}C isotopes

We also demonstrated the applicability of the SR microradiography for visualization of the fish scales fine structure with high–resolution (Fig 3). This non–destructive method can be used for study of the sample features prior the chemical mapping, line scanning or depth profiling.

Conclusion

Laser assisted techniques are suitable for elemental analysis in fish scales. Combination of different modes can give information about 2D or 3D distribution of elements in such heterogeneous and layered samples.

Line scan measurements confirmed the chemical composition of the upper layers of scales (hydroxyapatite), which contains evenly spaced elements, including calcium, phosphorus, sodium, magnesium, manganese and strontium. The measurements of depth profiles revealed the contents of the lower layers of scales and the collagen layer, which contains mainly carbon.

Acknowledgement

The work has been supported by Ministry of Education, Czech Republic on the frame of grants MSM0021622412 and MSM0021630508. J.K. acknowledges Central European Initiative (CEI) for CERES fellowship.

References

Ikoma T, Kobayashi H, Tanaka J, Walsh D, Mann S (2003) Microstructure, mechanical and biomimetic properties of fish scales from Pagrus major, Journal of Structural Biology, 142:327–333

Onozato H, Watabe N (1979) Studies on fish scale formation and resorption, Cell Tissue Research 201:409–422

Torres FG, Troncoso OP, Naka matsu J, et al. (2008) Characterization of nanocomposite laminate structure occurring in fish scales from Arapaima Giga, Materials science & Engineering C–biomimetic and supramolecular systems, 28:1276–1283

J Biochem Tech (2010) 2(5):S50-S51
ISSN: 0974-2328

Optimization of method for study of influence of fluoranthene on the cell

Ondrej Zitka, Ondrej Vodicka, Petr Babula, Ladislav Havel, Marie Kummerova, Miroslava Beklova, Rene Kizek*

Received: 25 October 2010 / Received in revised form: 13 August 2011, Accepted: 25 August 2011, Published: 25 October 2011
© Sevas Educational Society 2011

Abstract

Polycyclic aromatics hydrocarbones (PAHs) belong to the group of the most occurring pollutants of living environment. They are able to enter ecosystems and subsequently live organisms due to their lipophilic properties. Their influence on living organisms is well known especially in the case of animals and especially human, influence of PAHs on plants is not recognized and described for the present. For this purpose, we developed rapid method for separation and detection of fluoranthene and similar PAHS, such as pyrene and benzo[*a*]pyrene, in plant extracts. Method consists in extraction and detection, where high performance liquid chromatography (HPLC) system with UV detection was used.

Ondrej Zitka, Rene Kizek*

Department of Chemistry and Biochemistry, Faculty of Agronomy, Mendel University in Brno, Zemedelska 1, CZ-613 00 Brno, Czech Republic.

*Tel: +420 545 133 350, Fax: +420 545 212 044
E-mail: kizek@sci.muni.cz

Ladislav Havel

Department of Plant Biology, Faculty of Agronomy, Mendel University in Brno, Zemedelska 1, CZ-613 00 Brno, Czech Republic.

Ondrej Vodicka, Petr Babula

Department of Natural Drugs, Faculty of Pharmacy, University of Veterinary and Pharmaceutical Sciences, Palackeho 1-3, CZ-612 42 Brno, Czech Republic.

Miroslava Beklova

Department of Veterinary Ecology and Environmental Protection, Faculty of Veterinary Hygiene and Ecology, University of Veterinary and Pharmaceutical Sciences, Palackeho 1-3, CZ-612 42 Brno, Czech Republic.

Marie Kummerova.

Department of Experimental Biology, Faculty of Science, Masaryk University, Kotlarska 2, CZ-611 37 Brno, Czech Republic.

Keywords: Polycyclic aromatics hydrocarbonates, fluoranthene, BY-2 cells, reactive oxygen species.

Introduction

PAHs demonstrated mutagenic effect on many cell lines based on interferences with DNA. This fact is well evident especially after activation of PAHs by UV, when reactive forms – radicals originate. Ability of PAHs to generate reactive oxygenic species (ROS) is discussed too. ROS can subsequently oxidatively damage large scale of biomolecules and can serve as apoptotic signals. In area of plant physiology, knowledge about PAHs effects and their mechanisms of actions are still missing and are only limited. Moreover, induction of oxidative stress by several PAHs was demonstrated. Fluoranthene represents one of the PAHs models, which is generally used in *in vitro* as well as *in vivo* experiments. For analysis of fluoranthene and other non-polar compounds, HPLC-UV system is very helpful in this type of experiments (Kummerova et al. 2006; Wu et al. 1998).

Materials and Methods

Aim of this work consisted in developing of method for separation and detection of PAHs in tobacco cells treated by fluoranthene. HPLC-ED system consisted of two chromatographic pumps (Model 582, ESA Inc., Chelmsford, MA, working range 0.001-9.999 ml min^{-1}), chromatographic column with reverse phase Phenomenex Gemini NX C18 (100 × 2,0; 3 μm particles, Phenomenex, USA) and UV detector (Model 528, ESA, USA). Sample (20 μl) was injected by autosampler (Model 542, ESA, USA), which has thermostated space for column. All BY-2 cell samples were first destructed in liquid nitrogen and then extracted in acetonitrile. Samples of matrixes were then centrifuged at 14 000 G by time of 20 minutes. Supernatant was then directly analyzed by HPLC.

Results and discussion

We searched the best parameters and compromises in optimizing of separation method. Most suitable flow rate of mobile phase was determined as 0.3 ml min^{-1}. Mobile phase consisted of A: (acetic acid, 100 mM) and B (acetic acid, 100 mM, in methanol). Compounds of interest were eluted by linearly increasing gradient:

0-6 min (70% B), 6-10 min (100 % B), 10-13 min (100 % B). Chromatographic column

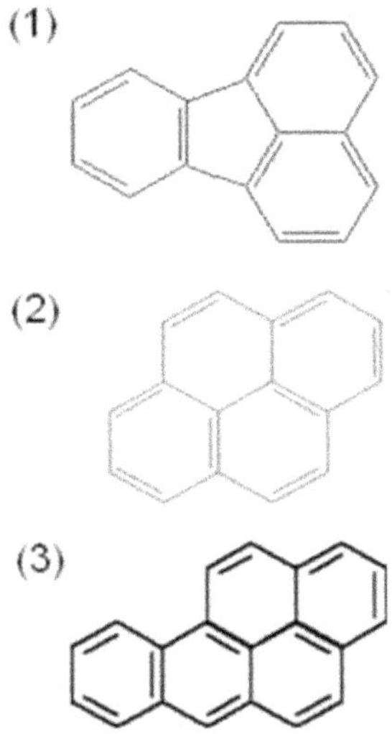

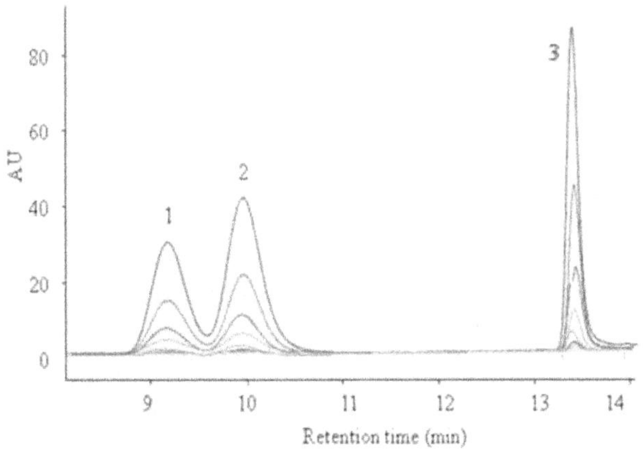

Figure 1: Structure of tree basicly studied PAH´s. (1) fluoranthene, (2) pyrene and (3) benz[a]pyrene

was thermostated to 25°C. Thanks to developing and optimizing of separation method, we were able to effectively separate non-polar compounds, as fluoranthene, pyrene, benzo[*a*]pyrene, naphtalene, anthracene, methyl-anthracene, benzylanthracene, triphenyl and coronen. All compounds were detected at one universal wavelength 275nm which was the most suitable and enough sensitive with regards to absorption maximum for all studied compounds without discrimination.

Figure 2: Chromatograms of calibration curve of three PAH´s fluoranthene (1), pyrene (2) and benz[a]pyrene.

The loose of resolution for peaks 1 and 2 due to usage of short column is shown in Fig.1. We found compromise of temperature and accurate adjusted isocratic ration of MF and thus we were able to avoid to bigger coelution. Anyway there is obvious effect of increasing of organic part of MF on peak 3 which has comparatively better resolution.

Conclusion

We achieved compromise in fast and robust separation of above-mentioned nonpolar compounds using short column with small

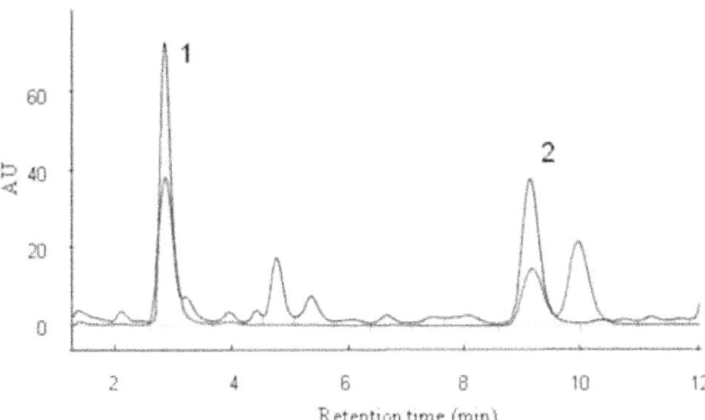

Figure 3: Comparison of standard (blue – line) and sample of extract of BY-2 cells treated by fluorantene (dark-line). Peak (1) internal standard-Toluene. Peak (2) fluorantene.

inner diameter. All of compounds of interest can be simultaneously determined in time interval shorter than 25 minutes including regeneration of column and equilibration of the system. The major benefits of this method are application during study of concentration of PAH´s in the plants after treatment. Due to simple preparation of the sample and or relatively fast separation and universal detection.this method is easy to use and robust for a high throughput data analysis.

Acknowledgements

The work has been supported by GA ČR 522/09/0239, MSMT 6215712402 and 1M06030.

References

Kummerova M, Krulova J, Zezulka S, Triska J (2006) Evaluation of fluoranthene phytotoxicity in pea plants by Hill reaction and chlorophyll fluorescence. Chemosphere 65(3):489-496

Wu PF, Chiang TA, Wang LF, Chang CS, Ko YC (1998) Nitro-polycyclic aromatic hydrocarbon contents of fumes from heated cooking oils and prevention of mutagenicity by catechin. Mutation Research-Fundamental and Molecular Mechanisms of Mutagenesis 403(1-2):29-34

J Biochem Tech (2010) 2(5):S52-S53
ISSN: 0974-2328

The photodegradation of polymers and small molecular materials applied in organic optoelectronic devices

Patricie Heinrichova, Petra Moziskova, Martin Sedina, Martin Weiter*

Received: 25 October 2010 / Received in revised form: 13 August 2011, Accepted: 25 August 2011, Published: 25 October 2011
© Sevas Educational Society 2011

Abstract

This contribution is focused on study of photo-degradation of a several photoconductive organic materials such as polymeric (high Tg–PPV – block copolymer of derivates of poly(p–phenylene–vinylene) and P3HT – poly(3–hexylthiophene–2,5–diyl) as small molecular weight material a derivate of diphenyl–diketopyrrolo–pyrrole – DPP 36 was used. These materials are used for construction of optoelectronic devices like organic solar cells, transistors, optical sensors and others. Photo-degradation processes were studied by optical characterization (UV–VIS spectroscopy,) and by analysis of photographs obtained by means of optical microscope.

Keywords: UV–VIS spectroscopy, polymers, photodegradation

Introduction

Opto-electrical devices use absorption of light for generation of charge carries. The photogeneration process competes with several photophysical processes as fluorescence, phosphorescence, thermal relaxation and others, which don't have destructive effect on material, and with chemical reactions initialized by light, which lead to decomposition of the organic functional materials. Rate of photo-degradation depends on many factors, but mainly on light intensity and oxygen concentration, sample thickness, multilayer construction, used solvent, etc. (Jorgensen et al. 2008).

In this work, rate of photo-degradation was studied, which is very important parameter for future application and lifetime assessment of organic materials in optoelectronics.

Patricie Heinrichova, Petra Moziskova, Martin Sedina, Martin Weiter*

Brno University of Technology, Faculty of Chemistry, Purky ňova 464/118, Brno 61200, Czech Republic

*Tel: +420 541 149 484, Fax: +420 541 211 697
E-mail: weiter@fch.vutbr.cz

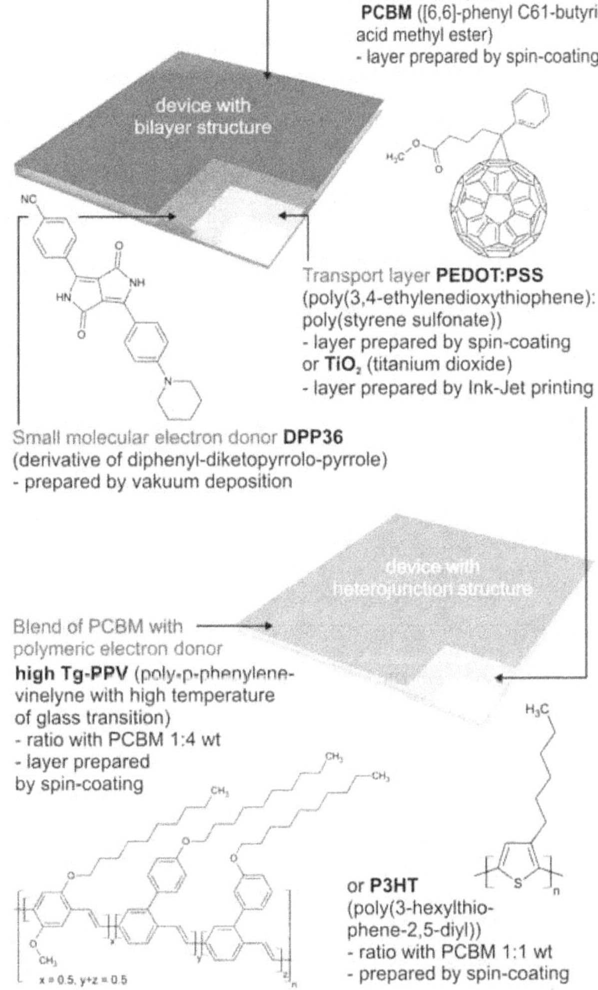

Figure 1: The scheme of the studied prepared samples and structures of used materials

Materials and Methods

Two polymers and one small molecular material were studied. For the experiment a π–conjugated block copolymer of derivatives of poly–*p*–phenylene–vinylene with high glass transition temperature (Tg–PPV) and P3HT (poly(3–hexylthiophene–2,5–diyl)) which were blended with PCBM ([6,6]–phenyl C61–butyric acid methyl ester) were chosen. Tg–PPV was mixed with PCBM in the ratio 1:4 w/w and P3HT in ratio 1:1 w/w. Solutions were prepared in chlorobenzene. Polymer layers were created employing spin–coating.

As a small molecular weight material the derivative of diphenyl–diketopyrrolo–pyrrole (here abbreviated as DPP 36, see Fig 1) was used. The layer was created by vacuum evaporation. Layer of PCBM was prepared by spin–coating onto the DPP36. In all cases these active materials were deposited on charge transporting layer of PEDOT:PSS (poly(3,4–ethylenedioxy–thiophene):poly(4–styren sulfonate)) or TiO_2 (titanium dioxide). As a substrate quartz glass was used. Assessed samples were prepared without encapsulation and at ambient air atmosphere.

Samples were irradiated in Qsun Xe test chamber; model Xe–1–B with outdoor filter. Temperature in test chamber was 45 °C. Degree of degradation of materials was studied by measuring of UV–VIS absorption spectra on Varian Cary 50 Spectrophotometer. Simultaneously appearance of surface of samples was studied by the optical microscope NIKON Eclipse E200 and documented by photographs.

Table 1: The resulting rate constants of the studied samples, where n is the order of reaction

Active layer	n	Rate constants (10^{-2} mol^{n}h^{-1})	
		/PEDOT	/TiO_2
Tg-PPV:PCBM (1:4)	1	$15,7 \pm 0,6$	$13,7 \pm 0,6$
P3HT:PCBM (1:1)	0	$2,85 \pm 0,09$	$2,066 \pm 0,003$
DPP36/PCBM	0	$0,231 \pm 0,011$	$1,16 \pm 0,02$

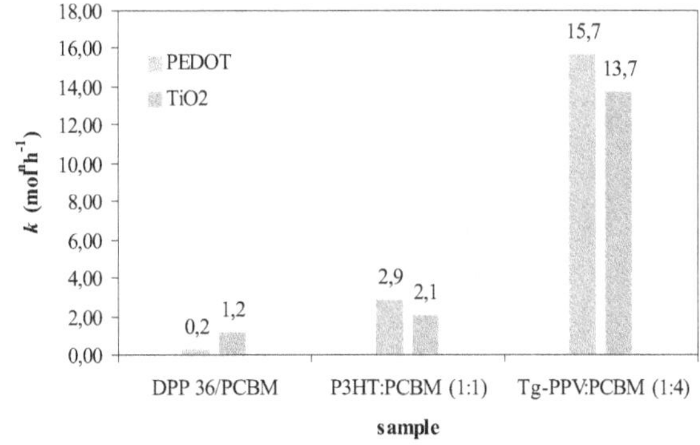

Figure 2: The diagram of the resulting rates constants of the studied samples.

Results and Discussion

These photo-degradation tests took 55 hours each. Kinetics of photo-degradation was obtained from the decay of characteristic absorbance maxima of UV–VIS spectra of researched electron donor materials. P3HT and DPP 36 presented zero order kinetic reaction while Tg–PPV showed more complex degradation kinetics, however for a first approach; it was possible to simplify it to the first order reaction.

Significant difference has been found between use of TiO_2 and PEDOT transport layers for DPP 36, when kinetic decay constant for PCBM/DPP 36/TiO_2 was found 5 times bigger as for PCBM/DPP 36/TiO_2. Type of transport layer does not play such an important role in photo-degradation processes of used polymers. The rate of photo-degradation of the low molecular DPP 36 was slower than in the polymers, while PCBM/DPP36 presented 4 times lower average rate constant when compared to P3HT:PCBM and 24 times lower if compared to Tg-PPV:PCBM respectively. Decomposition kinetics of Tg-PPV:PCBM was almost 6 times faster as P3HT:PCBM. The results are showed in diagram on Fig 2 and in Table 1.

Simultaneously, appearance of samples' surfaces was analyzed by means of the optical microscopy and documented. Expected color changes were observed during the whole experiment while the TiO_2 layer cracking was presented for the first four hours of the experiment. This cracking contributed to mechanical disintegration of DPP 36 layer. This fact naturally influenced trend of absorbance decay (disintegration of layer caused faster decay of absorbance with time of light irradiation of samples).

Conclusions

These photo-degradation assessments presented the enhanced stability of the studied DPP 36 dye with PCBM layer over the polymer blends Tg-PPV:PCBM (1:4) and P3HT:PCBM (1:1). Moreover, significant difference in stability between blends of P3HT and Tg-PPV was found.

The importance of charge transport layer selection was observed mainly for DPP 36 while in case of polymers the influence was negligible. Although the DPP 36 as a low molecular photoconductive material was found more photo stable against the the polymer materials, it showed lower mechanical endurance, because the TiO_2 cracking contributed to the disintegration of the DPP 36 layer, but not of the polymer layers. DPP 36 layer on PEDOT remained compact during the entire photo-degradation test.

Acknowledgement

The work has been supported by Grant Agency of the Czech Republic via project No. P205/10/2280.

References

Jorgensen M, Norrman K, Krebs FC (2008) Stability/degradation of polymer solar cells. Solar Energy Materials and Solar Cells 92(7):686-714

J Biochem Tech (2010) 2(5):S54-S55
ISSN: 0974-2328

Effect of heavy metals on the antioxidant activity of *Silene dioica*

Olga Krystofova, Gabriela Hlobilová, Eva Kramářová, Josef Zehnalek, Vojtech Adam, Rene Kizek*

Received: 25 October 2010 / Received in revised form: 13 August 2011, Accepted: 25 August 2011, Published: 25 October 2011
© Sevas Educational Society 2011

Abstract

This contribution is focused on study of photo-degradation of a several photoconductive organic materials such as polymeric (high Tg–PPV – block copolymer of derivates of poly(p–phenylene–vinylene) and P3HT – poly(3–hexylthiophene–2,5–diyl) as small molecular weight material a derivate of diphenyl–diketopyrrolo–pyrrole – DPP 36 was used. These materials are used for construction of optoelectronic devices like organic solar cells, transistors, optical sensors and others. Photo-degradation processes were studied by optical characterization (UV–VIS spectroscopy,) and by analysis of photographs obtained by means of optical microscope.

Keywords: UV–VIS spectroscopy, polymers, photodegradation

Introduction

Living organisms are constantly exposed to various health burdensome substances. Among these substances are also present heavy metals. These are traces of a part of the earth's crust and the environment is increasingly receiving the impact of human activity. Their effects on living organisms' produces a large number of biochemical reactions whose products may be free radicals, which in excess can cause oxidative stress. The protective system of the organism is based on the activity of specific enzymes (especially superoxid dismutase, glutathion peroxidase, catalase, glutathione reductase) as well as non-enzymatic compounds with antioxidant activity (L-ascorbic acid, glutathione). (Babula et al. 2008; Stratil et al. 2006; Krystofova et al. 2009)

Olga Krystofova, Josef Zehnalek, Vojtech Adam, Rene Kizek*

Department of Chemistry and Biochemistry, Department of Plant Biology Faculty of Agronomy, Mendel University in Brno, Zemedelska 1, CZ-613 00 Brno, Czech Republic

*Tel: +420 545 133 350, Fax: +420 545 212 044
E-mail: kizek@sci.muni.cz

Gabriela Hlobilová, Eva Kramářová

Department of Natural Drugs, Faculty of Pharmacy, University of Veterinary and Pharmaceutical Sciences, Palackeho 1-3, CZ-612 42 Brno, Czech Republic

Materials and methods

In our experiment, we presented *Silene dioica* suspension cultures of different concentrations of cadmium ions (0, 10, 50 and 100μM), and monitored its effect on antioxidant activity using DPPH methods, TEAC and FRAP. The DPPH• test is based on the ability of the stable 2,2-diphenyl-1-picrylhydrazyl free radical to react with hydrogen donors. The FRAP method (Ferric Reducing Antioxidant Power) is based on the reduction of complexes of 2,4,6-tripyridyl-s-triazine (TPTZ) with ferric chloride hexahydrate ($FeCl_3 \cdot 6H_2O$), which are almost colourless, and eventually slightly brownish. Results obtained using this method are usually recalculated to Trolox® concentration and are described as "Trolox® Equivalent Antioxidant Capacity" (TEAC). For chemically pure compounds, TEAC is defined as the micromolar concentration of Trolox® equivalents demonstrating the same antioxidant activity as a tested compound (at 1 mmol·L-1concentration).

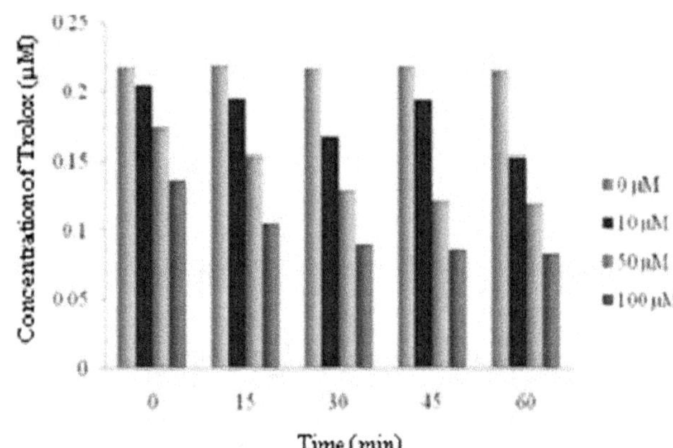

Figure 1: Diagram showing *Silene dioica* suspension cultures of different concentrations of cadmium ions (0, 10, 50 and 100μM) using DPPH and FRAP methods

Results and Discussion

The results obtained showed that in the case of DPPH and FRAP showed a suspension culture compared to the control of the reduced antioxidant activity, with a loss of activity was proportional to the concentration-acting metal (Fig 1). In the case of the TEAC method was observed opposite trend in activity than previous methods, namely the suspension cultures showed a higher antioxidant activity than the control for all the applied concentrations of heavy metals.

Conclusion

Environmental contamination by heavy metals is already known two-stroke problem. Unfortunately, the unintended consequences of contamination are still not entirely known. It is therefore necessary for their detection is of interest as their impact on living organisms metabolom to the contamination of the environment not easily identified, but also of arrest, or eliminate.

Acknowledgement

Financial support from the following grants TP 1/2010, GAČR 522/07/0692 and GAČR 204/09/H002 is highly acknowledged.

References

Babula P, V Adam et al (2008) Uncommon heavy metals, metalloids and their plant toxicity: a review. Environmental Chemistry Letters 6(4):189-213

Krystofova O, V Shestivska et al (2009) Sunflower Plants as Bioindicators of Environmental Pollution with Lead (II) Ions. Sensors 9(7):5040-5058

Stratil P, B Klejdus et al (2006) Determination of total content of phenolic compounds and their antioxidant activity in vegetables - Evaluation of spectrophotometric methods. Journal of Agricultural and Food Chemistry 54(3):607-616

J Biochem Tech (2010) 2(5):S56-S57
ISSN: 0974-2328

Effect of heavy metals ions on the antioxidant activity of Norway spruce tissue cultures

Olga Krystofova, Tereza Zizkova, Josef Zehnalek, Vojtech Adam, Karel Slaby, Ladislav Havel, Rene Kizek*

Received: 25 October 2010 / Received in revised form: 13 August 2011, Accepted: 25 August 2011, Published: 25 October 2011
© Sevas Educational Society 2011

Abstract

Heavy metal contaminations of land resources continue to be the focus of numerous environmental studies and attract a great deal of attention worldwide. Plants experience oxidative stress upon exposure to heavy metals that leads to cellular damage and disturbance of cellular ionic homeostasis. Our work was aimed at investigation of cadmium influence on Norway spruce explant culture and its effect on antioxidant activity.

Keywords: Norway spruce, heavy metals, antioxidant activity, cadmium

Introduction

Heavy metals still represent a group of dangerous pollutants, to which close attention is paid. However, all metals, especially cadmium are toxic at high concentration because of disrupting enzyme functions, replacing essential metals in pigments or producing reactive oxygen species. (Babula et al. 2008; Stratil et al. 2006; Supalkova et al. 2007)

Materials and Methods

Our work was aimed at investigation of influence of cadmium on early somatic embryos of Norway spruce. The plants were exposed to the metal ions concentrations of 0, 50, 250 and 500 µM for 11

Olga Krystofova, Josef Zehnalek, Vojtech Adam, Ladisla v Havel, Rene Kizek*

[1]Department of Chemistry and Biochemistry, Faculty of Agronomy, Mendel University in Brno, Zemedelska 1, CZ-613 00 Brno, Czech Republic

Tereza Zizkova, Karel Slaby

[2]Department of Plant Biology Faculty of Agronomy, Mendel University in Brno, Zemedelska 1, CZ-613 00 Brno, Czech Republic

*Tel: +420 545 133 350, Fax: +420 545 212 044
E-mail: kizek@sci.muni.cz

days. The Petri dishes with embryos were sampled at 4th, 6th, 8th, 10th and 11th day of experiment and subsequently analysed using electrochemistry. We focused our attention on observing of growth characteristics, the content of heavy metals in embryos and antioxidation activity.

Results and Discussion

In the first part of the experiment, a significant inhibitory effect at all concentrations of cadmium on growth of spruce tissue cultures was observed (Fig 1).

In the second part, we observed the influence of cadmium on antioxidant activity. To determine antioxidant activity the methods FRAP and DPPH was used. The obtained results show that group

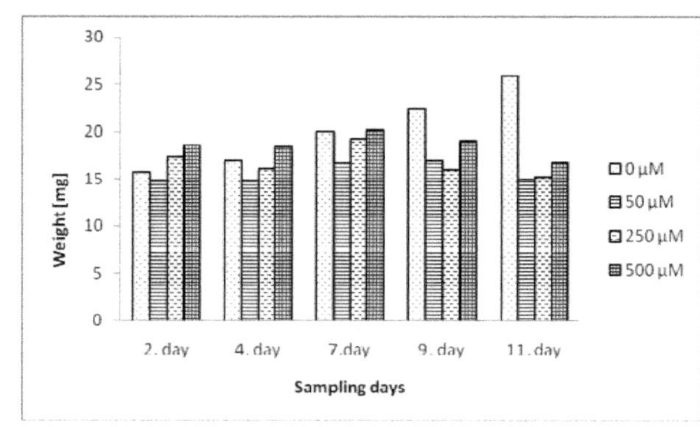

Figure 1: Average weight of tissue cultures of Norway spruce during the experiment

treated by cadmium compared to control group increased the antioxidant activity.

Conclusion

The obtained results show that heavy metals (in our case represented by cadmium) affect the metabolic cycles of living organisms and therefore it is important to pay our attention to this kind of research.

Acknowledgement

Financial support from the following grants GAČR 204/09/H002 and GAČR 522/07/0692 is highly acknowledged.

References

Babula P, V Adam et al (2008) Uncommon heavy metals, metalloids and their plant toxicity: a review. Environmental Chemistry Letters 6(4): 189-213.

Stratil P, B Klejdus et al (2006) Determination of total content of phenolic compounds and their antioxidant activity in vegetables - Evaluation of spectrophotometric methods. Journal of Agricultural and Food Chemistry 54(3): 607-616

Supalkova V, J Petrek et al (2007). Multi-instrumental investigation of affecting of early somatic embryos of spruce by cadmium(II) and lead (II) ions. Sensors 7(5): 743-759

J Biochem Tech (2010) 2(5):S58-S59
ISSN: 0974-2328

Study of microorganisms degrading PCB in vegetated contaminated soil

Veronika Kurzawova, Ondrej Uhlik, Martin Strohalm, Jan Lipov, Tomas Macek, Petr Stursa, Martina Mackova*

Received: 25 October 2010 / Received in revised form: 13 August 2011, Accepted: 25 August 2011, Published: 25 October 2011
© Sevas Educational Society 2011

Abstract

Removal of PCBs from contaminated soil is one of the challenges of environmental microbiology. In our study, we aimed to isolate, characterize and identify microorganisms from contaminated soil and to find out the plant effect on microbial diversity in the environment. Microorganisms were isolated by two ways, direct extraction and isolation after cultivation with biphenyl as a sole source of carbon. Isolated bacteria were biochemically characterized and the composition of ribosomal proteins in bacterial cells was determined by mass spectrometry MALDI-TOF. Bacteria with required properties were chosen and the *bphA* gene was amplified and detected. Bacteria with detected *bphA* gene were then identified by 16S rRNA sequence analyses.

Keywords: PCB, microbial diversity, *bphA* gene, 16S rRNA, MALDI-TOF.

Introduction

Polychlorinated biphenyls are organic compounds that were used in industry in the last century and during this time, large amounts of these compounds were released to the environment. Later on their toxicity was discovered and they use was stopped (Demnerova et al. 2005). Removing PCBs from contaminated soil using biological systems is inexpensive, easy and environmentally friendly process. Bacteria degrade PCBs by enzymes encoded by genes that are included in the biphenyl opern. The first gene of the biphenyl operon is the *bphA* gene encoding biphenyl dioxygenase (Abramowicz 1990). The amplification and detection of this gene is one of the way how to verify the degradative potential of bacteria.

Veronika Kurzawova, Ondrej Uhlik, Martin Strohalm, Jan Lipov, Tomas Macek, Petr Stursa and Martina Mackova*

Department of Biochemistry and Microbiology, Faculty of Food and Biochemical Technology, ICT Prague, Technická 3, Prague, 166 28

*Tel: +420 220 445 139, Fax: +420 224 355 167
E-mail: martina.mackova@vscht.cz

Materials and Methods

Samples were prepared from long term contaminated soil with four paralels vegetated by *Nicotiana tabacum* (tobacco) and four by *Solanum nigrum* (nightshade). Different bacterial species were inoculated to vegetated soil. The control sample was non-vegetated and with no bacteria added. Samples were kept in greenhouse for three months and then cultivable bacteria were isolated, characterized and identified. Bacteria were isolated by two ways, by direct extraction and after long term enrichment cultivation with biphenyl as a sole source of carbon and energy. Obtained isolates were characterized by biochemical tests, and the composition of ribosomal proteins was determined by mass spectrometry MALDI-TOF. The plant effect on microbial diversity was determined.

Results and Discussion

38 isolates were obtained after isolation by direct extraction from contaminated soil and 18 after cultivation with biphenyl. These isolates were biochemically and analytically characterized. Chosen isolates were analyzed by molecular-biological methods, *bphA* gene detection and 16SrDNA sequencing were performed in DNA of colonies growing on minimalm medium with biphenyl.

The experiments showed that plants significantly affect the microbial diversity in the rhizosphere. It was documented, that in tobacco rhizosphere different bacterial species than in nightshade rhizosphere were identified. More bacterial species able to degrade PCBs were detected in tobacco rhizosphere. These data correlate with findings published by Mackova et al. 2009, when tobacco was shown as the plant strongly supporting PCB degraders in studies carried out during ten years with different plant species. In our experiment only one bacterial species was isolated from both rhizospheres. The bacteria of the genus *Pseudomonas* were detected in all samples including the control sample.

Methods used for bacterial characterization and identification were compared. The fastest and the easiest analytical method is MS MALDI-TOF, but the proper and expensive instrument and specific database are needed. The 16S rRNA gene sequence analysis is still the most accurate method for bacterial identification and for studying microbial diversity in the environment.

Conclusion

Comparison of the methods for the characterization and identification of bacteria is one of the aim of environmental microbiology. In our study we cultivated two plant species in PCB contaminated soil and after several months of growth we compared microbial diversity in rhizosphere of both plants after straight extraction and enriched cultivation with several passages with biphenyl. After stright extraction 38 isolates were identified, 11 of them possesed *bphA* gene, enrichment isolation gave 18 isolates with 8 containing *bphA* gene. Described microorganisms obtained from these experiments can be used for bioremediation. Spesies identified were classified as genera *Pseudomonas, Achromobacter* and *Ochrobactrum*. After enrichment isolation *Pseudomonas mendocina* and *Pseudomonas alcaliphila* were identified. Other species were isolated after straight extraction from the soil. In rhizosphere of tobacco 7 bacterial species were identified in nightshade rhizospere only 4 species were detected. Analysis of PCB removal showed that the lowest concentration of PCB was measured in soil vegetated by tobacco with augmented bacteria of genus *Pseudomonas*. In this case almost 53% of original PCB concentration were removed.

Acknowledgement

This work was supported by the grants GACR 525/09/1058 a MSM 6046137305, MSMT ME 09024.

References

Abramowicz DA (1990) Aerobic and anaerobic biodegradation of PCBs- a review. Critical Reviews in Biotechnology 10(3):241-249

Demnerova K, Mackova M et al. (2005) Two approaches to biological decontamination of groundwater and soil polluted by aromatics - characterization of microbial populations. International Microbiology 8(3):205-211

Mackova M, Prouzova P et al (2009) Phyto/rhizoremediation studies using long-term PCB-contaminated soil. Environmental Science and Pollution Research 16:817-829

J Biochem Tech (2010) 2(5):S60-S61
ISSN: 0974-2328

Validation of duckweed microbiological test for assessing hazardous substances

Ivana Soukupova*, Miroslava Beklova

Received: 25 October 2010 / Received in revised form: 13 August 2011, Accepted: 25 August 2011, Published: 25 October 2011
© Sevas Educational Society 2011

Abstract

For this study, we used modified EN ISO 20079. The modification consists in application of polystyrene macroplates, with the advantage of requiring low sample volumes for the test (10 ml). *Lemna minor* was cultivated as a monoculture by various doses of chemicals for seven days. The aim of our work was to compare the acute toxicity ($168hEC_{50}$) obtained from conventional testing (100 ml) and from microbiological tests (10 ml) using reference toxicants: potassium chloride and 3.5 – dichlorphenol. The resulting value $168hEC_{50}$ for potassium chloride using conventional test was 9.78 g.l^{-1} and for 3.5 – dichlorphenol it was 5.71 mg.l^{-1}. The resulting values from microbiological tests were 8.69 g.l^{-1} and 4.24 mg.l^{-1}, respectively. The comparison of measured values of acute toxicity from conventional tests and microbiological tests indicates that microbiological tests are an appropriate alternative to today commonly used ecotoxicological biotests, for measuring and assessing of biological effects of toxic substances in the aquatic environment.

Keywords: duckweed (*Lemna minor*), microbiological tests, potassium chloride, 3.5 – dichlorphenol

Introduction

Due to the enormous number of potentially polluting substances contained in waste waters from municipal and environmental sources, here grows a necessity of providing the information about water quality. A chemical-specific approach is insufficient to provide the complex information about water quality. Therefore, it is essential to use biological test systems with living cells or organisms that give a global response to the quantum of micropollutants present in the sample.

Ivana Soukupova, Miroslava Beklova*

Department of Veterinary Ecology and Environmental Protection, Faculty of Hygiene and Ecology, University of Veterinary and Pharmaceutical Sciences Brno, Palackeho 1/3, 612 42 Brno

*Tel: +420 541 562 652, Fax: +420 541 562 657
E-mail: beklovam@vfu.cz

Settled life style of plants makes them an organism constantly exposed to the pollution. Also plants are the major nutriment source for nearly all higher organisms and as such play an active role in transferring contaminants to higher trophic levels (Radic et al. 2010).

Duckweed (*Lemna minor*) is used in quality studies to monitor heavy metals and other aquatic pollutants. The plants posses physiological properties (small size, rapid growth between pH 5 – 9, and vegetative propagation), that make duckweed an ideal test system. Among the developmental parameters, the most commonly assessed in ecotoxicological test systems, are growth parameters (Wang and Freemark 1995).

In EN ISO 20079 Water quality – Determination of the toxic effect of water constituents and waste water on duckweed (*Lemna minor*) – Duckweed growth inhibition test are tested 100 ml sample volumes. Using microvolumes (10 ml) in microbiological tests can be a good tool to include in a battery of tests for phytotoxicity screening of a wide range of chemicals and environmental samples, with the advantage of allowing large numbers of samples to be tested, and generating low volumes of waste (Paixao et al. 2008).

The organisms commonly employed in microbiological tests are bacteria, protozoa, invertebrates, fish and tissue cultures etc. But standard microbiotest using duckweed as a test organism has not been created yet. The aim of this work is to validate a miniaturized duckweed test using macrotitration plates as a suitable alternative to conventional ecotoxicological biotests.

Materials and methods

Growth inhibition tests, with duckweed, were performed according to the standard EN ISO 20079. The 7-day toxicity test conducted at $24 \pm 2^{\circ}C$ in test vessels containing a minimum of 100 ml of test solution and 3-frons plants. Our suggested modification requires lower volumes of test solutions and reduced amount of plants. We employed a static test, where the test solutions are not renewed during the test. For the comparison of sensitivity and validation of the accuracy and reliability of the results of this method were performed examinations with the reference toxicants.

We investigated the effect of various doses of two defined referent toxicants: potassium chloride (3; 4; 7; 10 and 15 g.l^{-1}) and 3.5 – dichlorphenol (1.5; 2.1; 3; 4.2 and 5.9 mg.l^{-1}) on *Lemna minor*.

Plant material was obtained from a culture collection of the ecotoxicological laboratory of the University of Veterinary and Pharmaceutical Sciences Brno and was adapted for the test. The stock culture of duckweed is cultivated in SIS medium so we also used SIS medium with pH 6.5 for preparation of the concentration row (OECD 2004). In the situation when the validation requirements are fulfilled it is allowed using e.g. SIS medium (EN ISO 20079 2007).

We tested samples of 100 ml volume (conventional test: 150 ml beaker) and of 10 ml volume (Microbiological test) in the same time. Polystyrene macroplates used for the microbiological test consist of six dimples of maximum volume 15 ml with flattened bottom and with the cover. Test vessels in conventional test were covered by the foil to minimize evaporation and accidental contamination. All applied vessels avoided shadowing or changes in the spectral characteristics of light.

In 100 ml volume the initial number of fronds was nine. In microbiotest the number was reduced to five. The vessels with referent toxicants in three replicates, and the control in six replicates, were incubated during seven days under a continuous warm fluorescent lightning and with the temperature of 24±2°C.

For the test to be valid, the doubling time of frond number in the control must be less than 2.5 days, corresponding to approximately a seven-fold increase in seven days. The toxic influence of reference toxicants was evaluated on a basis of growth inhibition expressed as number of fronds and comparison of growth rates (Fig. 1 and 2) (EN ISO 20079 2007).

Formula 1:

$$\mu_{i\text{-}j} = \ln (N_j) - \ln (N_i) / t_{j\text{-}i}$$

Average specific growth rate
$\mu_{i\text{-}j}$: average specific growth rate from moment time i to j
N_i : number of fronds observed in the test or control vessel at time i (the end of the test)
N_j : number of leaves observed in the test or control vessel at time j (the start of the test)
t_i : moment time for the start of the period
t_j : moment time for the end of the period

Formula 2:

$$\% \ I_r = [(\mu_c - \mu_T) / \mu_c] \times 100$$

Percent inhibition of growth rate
$\% \ I_r$: percent inhibition in average specific growth rate
μ_c : mean value for μ in the control
μ_T : mean value for μ in the treatment group

Results and discussion

Our conventional tests and also the microbiological tests were able to fulfill the validity requirements. EN ISO 20079 specifies ranges of resulting 168hEC$_{50}$ values for both toxicants. The values of 168hEC$_{50}$ for potassium chloride (using APHA medium) has to lie in the range of 2.2 – 3.8 mg.l^{-1} and for 3.5 – dichlorphenol (modified Steinberg medium) in 5.5 – 10.0 g.l^{-1}. But the values of acute toxicity for both referent toxicants using SIS media are not known. The resulting values of 168hEC$_{50}$ for potassium chloride were corresponding with the declared range, for conventional test it was 9.78 g.l^{-1} and for microbiotest it was 8.69 g.l^{-1}. For 3.5 – dichlorphenol resulting values

of 168hEC$_{50}$ (5.71 mg.l^{-1} and 4.24 mg.l^{-1} respectively) were higher in both cases of testing (Table 1).

Table 1: Comparison of the resulting values of 168hEC$_{50}$ for potassium chloride and 3.5 – dichlorphenol

168hEC$_{50}$	KCl	3.5 - dichlorphenol
Conventional test	9.78 g.l^{-1} 95% interval of reliability = 9.16 – 10.40	5.71 mg.l^{-1} 95% interval of reliability = 5.52 – 5.90
Microbiotest	8.69 g.l^{-1} 95% interval of reliability = 8.18 – 9.20	4.24 mg.l^{-1} 95% interval of reliability = 4.15 – 4.33

Conclusion

From the resulting values of 168 hours lasting growth inhibition test using water bioindicator *Lemna minor* (168hEC$_{50}$) for two referent toxicants we can observe a good correlation between conventional (100 ml) test and microbiotest (10 ml). In our microbiotest we also carried out the validity requirements. So it can be seen that microbiological tests for assessing toxic effect of chemicals or other hazardous substances are a suitable alternative to commonly used ecotoxicological biotests.

Acknowledgements

This research was supported by the Ministry of Education, Youth and Sports of the Czech Republic (MSMT 6215712402) and FRVS 2691/2010.

References

EN ISO 20079 (2007) Water quality – Determination of the toxic effect of water constituents and waste water on duckweed (*Lemna minor*) – Duckweed growth inhibition test.

Paixao, S, M Silva L. et al. (2008) Performance of a miniaturized algal bioassay in phytotoxicity screening. Ecotoxicology 17:165-171

OECD Guidelines for the Testing of Chemicals Revised Proposal for a New Guideline 221(2004): *Lemna sp.* Growth Inhibition Test 221

Radic, S, Stipanicev D et al (2010) Ecotoxicological assessment of industrial effluent using duckweed (*Lemna minor L.*) as a test organism. Ecotoxicology 19(1):216-222

Wang W, Freemark K (1995) The use of plants for environmental monitoring and assessment. Ecotoxicology and environmental safety 30:289-301

J Biochem Tech (2010) 2(5):S62-S63
ISSN: 0974-2328

Adsorption studies of methyl blue and Cu^{2+} ions on lignite pellets

Petra Businova, Miloslav Pekar*

Received: 25 October 2010 / Received in revised form: 13 August 2011, Accepted: 25 August 2011, Published: 25 October 2011
© Sevas Educational Society 2011

Abstract

In recent years, possibilities of lignite utilization in various non–energy applications, especially in sorption processes, have been investigated. This paper reports on the ability of lignite pellets to adsorb methylene blue or copper ion (Cu^{2+}) from aqueous solution and motor oil from solid surface.

Keywords: Lignite pellets, sorption, motor oil, methylene blue, copper ion

Introduction

Among the various treatment methods, adsorption has been found to be an efficient and economic process to remove various pollutants. The most commonly used sorbent is activated carbon. It has a high adsorption capacity, but its high cost and troubles with regeneration limit its applications. Therefore, many non–conventional natural and locally available materials have been investigated as low–cost adsorbents (Babel and Kurniawan, 2003; Crini, 2006). This paper discusses the possibilities of using lignite pellets as a sorbent for removal of organic or inorganic substances from aqueous environment and oil spills from solid surface.

Materials and methods

Lignite pellets prepared by extrusion using a chopper (see Businova, 2007) were investigated as sorbents. Their composition can be seen in Tab. 1. An aqueous solution of methylene blue with an initial concentration of 1 g/L, aqueous solutions of $CuCl_2.2H_2O$, $CuSO_4.5H_2O$ and $Cu(NO_3)_2.3H_2O$ with an initial concentration of 0.1mol/L and motor oil were used as adsorbates. In the case of the dye and copper (II) compounds, 0.5 g of a sorbent was shaken with 50 mL of a solution and time of adsorption was 24 hours. An

Petra Businova, Miloslav Pekar*

Brno University of Technology, Faculty of Chemistry, Institute o f Physical and Applied Chemistry, Purkyňova 118, 612 00 Brno, Czech Republic

*Tel: +420 541 149 330, Fax: +420 541 211 697
E-mail: pekar@fch.vutbr.cz

adsorbate removal in a solution was determined using UV–VIS spectrophotometer. Residual concentration of particular sorbate in solution was calculated from absorbance at the wavelength of the absorption maximum. These wavelengths for methylene blue and copper were 664 and 810 nm, respectively. Sorption test using motor oil as an adsorbate was conducted in Petri dishes with 2 g of oil and 1 g of pellets. Contact time was 80 minutes. Adsorbed amount of oil was obtained from an increase in weight of the sorbent during the experiment.

Results and Discussion

Adsorption of three different adsorbates on lignite pellets was investigated. The amount of adsorbate sorbed per gram of pellets in 24 hours in the case of methylene blue and Cu^{2+} and in 80 minutes in the case of motor oil was determined. Furthermore, percentage removal of a dye and Cu^{2+} in 24 hours was calculated. The results can be seen in Fig. 1 and 2. The results show that all tested samples of pellets were able to adsorb both dye and copper ion from aqueous solution as well as oil from solid surface, although with different efficiency. Composition of pellets as well as the kind of adsorbate had effect on the resulting adsorbed amount. The order of affinity of lignite pellets towards adsorbates was found to be motor oil >> methylene blue > Cu^{2+}. The values of adsorbed amount were in the

Table 1: Composition of used pellets

Pellets	Binder	Composition (wt %) lignite + binder + water
1	Sokrat	83 + 4 + 13
2	Sokrat + molasses	82 + 6 + 3 +9
3	molasses	83 + 4 + 13
4	molasses	85 + 4 + 11
5	molasses	90 + 4 + 6
6	limestone + molasses + Sokrat	74 + 3 + 4 + 6 + 13
7	water–glass	79 + 6 + 15
8	water–glass + molasses	78 + 3 + 4 + 15
16e	Solvarin AP	55 + 6 + 39

range of 144–230 mg/g for motor oil, 46–99 mg/g for methylene blue and 14–58 mg/g for Cu^{2+}. In addition, effect of anions on copper ion adsorption was observed. Adsorption of Cu^{2+} in the presence of different anions decreased in the order $NO_3^- > SO_4^- > Cl^-$. The highest adsorbed amount of Cu^{2+} was 58 mg/g in the presence of nitrate, 42 mg/g in presence of sulfate and 28 mg/g in presence of chloride.

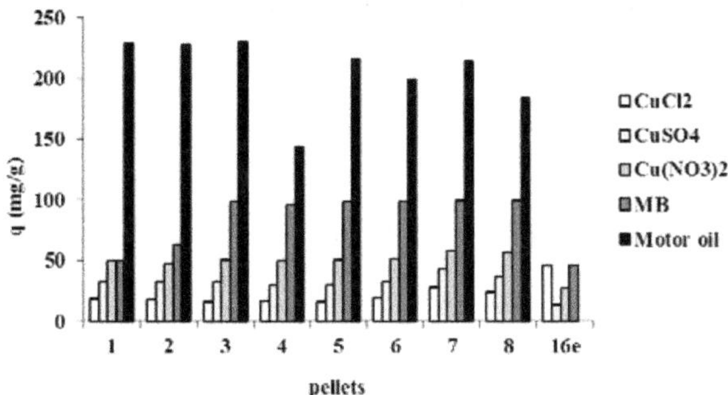

Figure 1: Adsorbate sorbed per gram of pellets

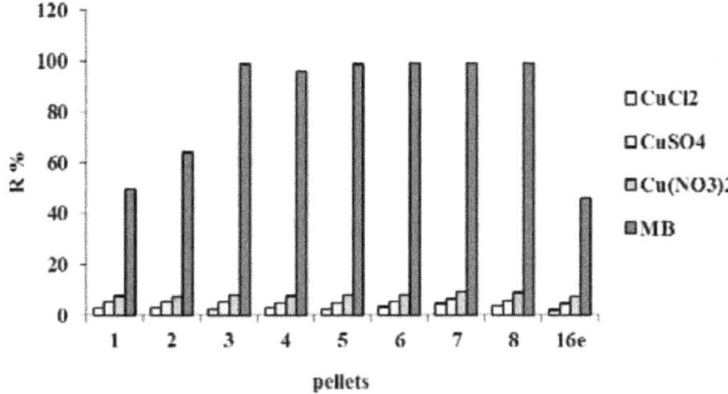

Figure 2: The percentage removal of Cu^{2+} and methylene blue from solution

Conclusion

The ability of lignite pellets to adsorb methylene blue, Cu^{2+} and motor oil was investigated. It was observed that all tested samples of pellets are able to adsorb all studied adsorbates, although with different efficiency. Adsorption efficiency can by influenced by composition of pellets.

Acknowledgement

This work was supported by the Czech government funding – Ministry of Education, project Nr. MSM0021630501.

References

Babel S, Kurniawan TA (2003) Low-cost adsorbents for heavy metals uptake from contaminated water: a review. Journal of Hazardous Materials 97(1-3):219-243

Businova P, Pekar M (2007) In 11th Conference on Environment and Mineral Processing, Part I. VŠB–TU Ostrava 319–323

Crini G (2006) Non-conventional low-cost adsorbents for dye removal: A review. Bioresour ce Technology 97(9):1061-1085

J Biochem Tech (2010) 2(5):S64-S65
ISSN: 0974-2328

Influence cadmium ions on the synthesis of thiol compounds for flax

Olga Krystofova, Václav Diopan, Jiří Baloun, Vojtech Adam, Rene Kizek*

Received: 25 October 2010 / Received in revised form: 13 August 2011, Accepted: 25 August 2011, Published: 25 October 2011
© Sevas Educational Society 2011

Abstract

Evaluation of the effectiveness of phytoremediation technologies is very difficult. One way to quickly and inexpensively identify phytoremediation potential of plants is found easily detectable marker. In our study, we examined the content of thiol compounds in plants, of Flax effects of various concentrations of cadmium ions.

Keywords: Flax, heavy metals, thiols, cadmium

Introduction

Phytoremediation is the direct use of living green plants to degrade, contain, or tender harmless various environmental contaminants, including recalcitrant organic compounds or heavy metals. The methods involved include phytoextraction, direct phytodegradation and rhizofiltration. Plants can remove organic and also inorganic compounds including heavy metals like mercury, cadmium, copper, lead, zinc etc. Organic compounds can be degraded or transformed by different enzymic systems like cytochromes P450, peroxidases, glutathione-transferases. Plants can also interact with rhizosphere degradation microorganisms by excreting of the root exudates. (Macek et al. 2008; Macek et al. 2007; Zehnalek et al. 2004). Electrochemical techniques are highly sensitized tool for the detection of peptides and proteins rich in thiol group applying to the response of the organism in the presence of heavy metals. Is certainly possible to argue, that these techniques do not have their place alongside modern analytic tools, such as mass spectrometry, electrophoresis with laser-induced fluorescence, but the concentration of these techniques, instead of their rivalry seems to be the most appropriate solution.

Olga Krystofova, Václav Diopan, Jiří Baloun, Vojtech Adam, Rene Kizek*

Department of Chemistry and Biochemistry, Department of Plant Biology Faculty of Agronomy, Mendel University in Brno, Zemedelska 1, CZ-613 00 Brno, Czech Republic

*Tel: +420 545 133 350, Fax: +420 545 212 044
 E-mail: kizek@sci.muni.cz

Materials and Methods

In our work, we focused our attention on different varieties of flax and their ability to uptake cadmium ions in different concentration and thus provide its use in phytoremediation technologies. Level of thiol compounds (glutathione and phytochelatins$_2$) and the total content of sulfur-rich compounds is one of the metabolomic indicators allowing the assessment of phytoremediation potential of flax. High-performance liquid chromatography with electrochemical detection was used for total content of thiol determination.

Results and Discussion

Analysis showed that Flax plants exposed to 80 µM concentration of cadmium ions, compared with control synthesize up to 40x more thiol compounds (Fig.1). Since thiol compounds play an important role in the detoxification of heavy metals in plants, based on the data obtained is attributable to the variety with the highest content of thiol compounds the highest phytoremediation potential.

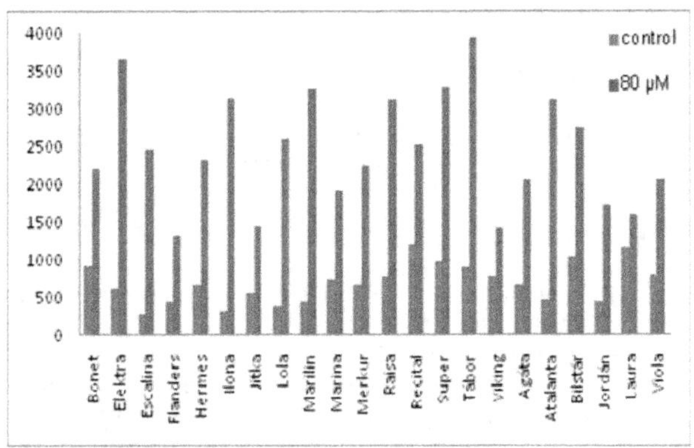

Figure 1: The total content of thiol compounds in different varieties of Flax exposed to cadmium ions exposure compared with control

Conclusion

Know the phytoremediation potential of plants is an important criterion for further experimental work and for their effective realization in practice. Content of thiol compounds could be one major marker, but also a marker of the degree of contamination of the site.

Acknowledgement

Financial support from the following grants GA ČR 522/07/0692 and 1M06030 is highly acknowledged.

References

Macek T, Kotrba P et al (2008) Novel roles for genetically modified plants in environmental protection. Trends Biotechnol 26(3):146-152

Macek T, Rezek J et al (2007) Phytoremediation Lis Cukrov Repar 123(9-10):312-314

Zehnalek J, Vacek J et al (2004) Application of higher plants in phytoremediation of heavy metals. Listy Cukrov 120(5-6):220-221

J Biochem Tech (2010) 2(5):S66-S67
ISSN: 0974-2328

Membrane-assisted fabrication of copper nanostructures and their characterization

Katerina Klosova, Libuse Trnkova*, Ota Salyk

Received: 25 October 2010 / Received in revised form: 13 August 2011, Accepted: 25 August 2011, Published: 25 October 2011
© Sevas Educational Society 2011

Abstract

The aim of the present study is to prepare new copper nanostructured electrodes by copper electroplating in porous nonconductive membranes and to examine them by scanning electron microscopy (SEM) and electrochemical impedance spectroscopy (EIS).

Keywords: Nanomaterials, copper nanostructures, electron microscopy, electrochemical impedance spectroscopy

Introduction

Nanomaterials and nanotechnology have received considerable attention recently due to the potential applications and properties that differ from those of bulk materials. For example, biologically important species can be immobilized on the nanostructures and also the area of the nanostructured electrode is larger in comparison to the flat electrode. Nanostructured surfaces seem to be promising in the field of sensing and biosensing as well as fuel cells, electronic elements, magnetic storage media, microsensors, optical devices, etc. (Hubalek et al. 2007; Metzger et al. 2000; Ramanathan et al. 2006). One of the most effective ways to produce dense arrays of metal nanostructures is metal electrodeposition into nanoporous nonconductive membrane.

Materials and methods

The membranes were composed of Al_2O_3 and vapour deposited with copper on one side in order to provide an ohmic contact and to cover the nanopores. During copper electroplating (from 0.8M $CuSO_4$ and

Katerina Klosova, Libuse Trnkova*

Department of Chemistry, Masaryk University, Kotlarska 2, 611 37 Brno, Czech Republic

*Tel: +420 549 497 754, Fax: +420 541 211 214
E-mail: libuse@chemi.muni.cz

Ota Salyk

Faculty of Chemistry, Brno University of Technology, Purky ňova 118, 612 00 Brno, Czech Republic

0.4M H_2SO_4 with current density 60 mA/cm^2, anode Cu wire, temperature 22°C), copper ions in a solution are attracted to the metal film at the bottom of the nanopores where they deposit in the form of solid metal. After filling the nanopores with copper, nanoporous membrane is dissolved in NaOH and the nanostructured surface is obtained (Fig 1).

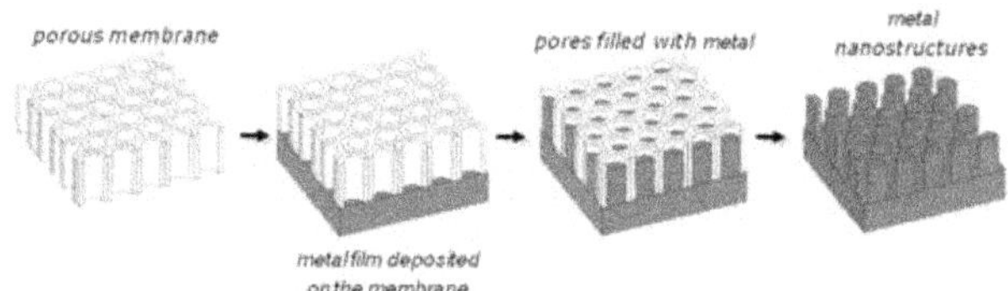

Figure 1: Illustration of copper nanostructure fabrication

The created copper nanostructures were examined using scanning electron microscopy (SEM) and electrochemical impedance spectroscopy (EIS). The conditions, under which majority of EIS experiments were performed, are summarized in Table 1.

Table 1: EIS experiment

electrolyte	1 mM KCl
frequency range	1MHz - 1kHz
potential	0 V
amplitude	0,01 V
number of frequencies	60
frequency step	logarithmic
wave type	single sine
auxiliary electrode	Cu
reference electrode	Ag/AgCl/3MKCl
temperature	22°C
device	Autolab 302N (EcoChemie)
software	Nova 1.5

Results and Discussion

An example of real copper nanostructures, formed by electroplating into membrane, is in Fig 2. Apart from SEM, the copper nanostructures were also characterized by EIS (Barsoukov 2005). The impedance has been carried out before and after electrodeposition into membrane, as well as after dissolution of the membrane. The EIS exhibited two semicircles (one in the high frequency region and one in the low frequency region) if the nanoporous membrane was present. In the case of EIS measured after dissolution of the membrane, the semicircle in the low frequency region was usually replaced with a straight line. An example of Nyquist plot is in Fig 3. The two semicircles in the Nyquist plot, obtained from the measurements with porous membranes, have been simulated using software Nova 1.5. The proposed equivalent circuit consists of one RC element, in which the capacitance C is usually replaced with the constant phase element Q, and the second part of the circuit is composed of another Q in parallel with finite Warburg impedance O. After dissolution of the membrane, the electrode behaves as an imperfect electrode coating.

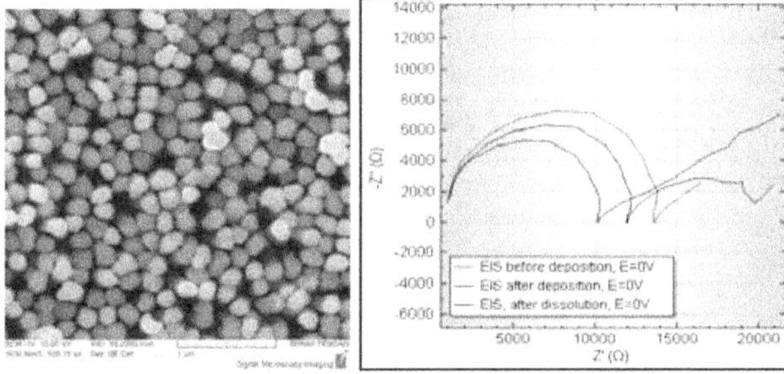

Figure 2: An example of copper   Figure 3: EIS measured with copper
auxiliary electrode

Conclusion

Membrane-assisted fabrication of copper nanostructures was used for the preparation of an array of nanostructures on a vapour deposited film. These nanostructures were examined by SEM and EIS. An equivalent electrical circuit for measured impedance spectra has been proposed.

Acknowledgement

The work has been supported by the following grants: INCHEMBIOL (MSM0021622412) and BIO-ANAL-MED (LC06035) from the Ministry of Education, Youth and Sports of the Czech Republic. The SEM analyses were provided by Tescan a.s.

References

Barsoukov E, Macdonnald JR (2005) Impedance Spectroscopy Theory, Experiment and Applications, 2nd ed., John Wiley & Sons, Inc

Hubalek J, Hradecky J, Adam V, Krystofova O, Huska D et al. (2007) Spectrometric and voltammetric analysis of urease - nickel nanoelectrode as an electrochemical sensor. Sensors 7(7):1238-1255

Metzger RM, Konovalov VV, Sun M, Xu T, Zangari G et al. (2000) Magnetic nanowires in hexagonally ordered pores of alumina. Ieee Transactions on Magnetics 36(1):30-35

Ramanathan S, Patibandla S, Bandyopadhyay S, Edwards JD, Anderson J (2006) Fluorescence and infrared spectroscopy of electrochemically self assembled ZnO nanowires: evidence of the quantum confined Stark effect. Journal of Materials Science-Materials in Electronics 17(9):651-655

J Biochem Tech (2010) 2(5):S68-S69
ISSN: 0974-2328

Fabrication of copper nanoparticles based screen–printed electrodes for electrochemical analysis

Jana Chomoucka*, Jan Prasek, Libuse Trnkova and Jaromír Hubalek

Received: 25 October 2010 / Received in revised form: 13 August 2011, Accepted: 25 August 2011, Published: 25 October 2011
© Sevas Educational Society 2011

Abstract

This paper is focused on the preparation and characterization of Cu_2O nanoparticles via simple wet chemical route. Samples were characterized by SEM and XRD. These nanoparticles were used for fabrication of screen–printed working electrodes for electrochemical detection of purine bases in DNA.

Keywords: Cu_2O nanoparticles, screen-printed electrode, cyclic voltammetry

Introduction

Cuprous oxide (Cu_2O) is a p–type metal oxide semiconductor with a direct band gap of 2.0–2.2 eV. It has attracted increasing interest due to its promising application in magnetic devices, solar energy conversion and catalysts (Huang et al. 2009; Zahmakiran et al. 2009).

New types of solid electrodes are necessary for small device technologies contrary to standard electrochemical analysis, where mercury drop electrodes are commonly used. The performance of solid electrode is determined by its surface modifications to make it sensitive and selective towards a certain analyte, to obtain either chemically or biochemically modified electrodes (Pravda et al. 2001). Solid electrodes can be fabricated by thick–film technology (TFT) process. The advantage of TFT is its flexibility, low production costs, good reproducibility and good electrical and mechanical properties of electrodes.

Jana Chomoucka*, Jan Prasek, Jaromír Hubalek

Department of Microelectronics, Faculty of Electrical Engineering and Communication, Brno University of Technology, Technick a 3058/10, CZ–616 00 Brno, Czech Republic

Ph: +420 541 146 163, Fax: +420 541 146 298
E-mail: chomoucka@feec.vutbr.cz

Libuse Trnkova

Department of Chemistry, Faculty of Science, Masaryk University, Kamenice 5, CZ–625 00 Brno, Czech Republic

Materials and Methods

The preparation method of Cu_2O/CuO nanoparticles is based on the procedure reported in (Luo et al., 2008). Nanoparticles were prepared by two–step synthesis. Copper (II) acetate reacts with sodium borohydride in three–phase system (water – N,N–dimethylformamide – cetyltrimethylammonium bromide) at 70°C to produce Cu_2O precursor. Then NaOH solution was added and the precursor was dried to produce Cu_2O/CuO nanoparticles. Prepared nanoparticles were characterized by SEM and XRD.

Two kinds of working electrodes were fabricated using standard TFT process – cermet and polymer. Cu_2O/CuO nanoparticles were well homogenised with 3 wt% of glass frit (in the case of cermet electrode) and suitable vehicle or polymer binder, respectively. Prepared pastes were screen–printed on an alumina substrate and fired. TFT electrodes were made only in the form of working electrode that is commonly a part of three-electrode system.

Electrochemical measurements were performed with AUTOLAB PGS30 Analyzer (EcoChemie, Netherlands) connected to VA-Stand 663 (Metrohm, Switzerland), using a standard cell. A three–electrode system was used, Cu_2O electrode was employed as the working electrode, an Ag/AgCl/3M KCl electrode served as the reference electrode and Pt electrode was used as the auxiliary electrode. Cyclic voltammetry (CV) were carried out in the presence of 0,2 M acetate buffer pH 5.0 and in the presence of 1.10^{-4} M adenine. CV parameters: scan rate 100 mV/s, potential range -0.5 to 0.5 V.

Results and Discussion

According to the XRD measurement, we prepared Cu_2O/CuO nanoparticles consisted of 60 % CuO and 40 % Cu_2O. In Figure 1 the differences between fired and non-fired cermet electrodes can be observed. In case of non-fired electrodes the Cu_2O/CuO nanoparticles were covered by vehicle whereas in the case of fired cermet electrode, the organic components were burned out and nanoparticles became naked. In the case of polymer electrode, the nanoparticles remained covered with polymer even fired, so the electrode active area remains minimal (Fig 2).

Figure 1: The SEM images of screen-printed cermet paste with Cu_2O/CuO particles before (left) and after (right) firing process.

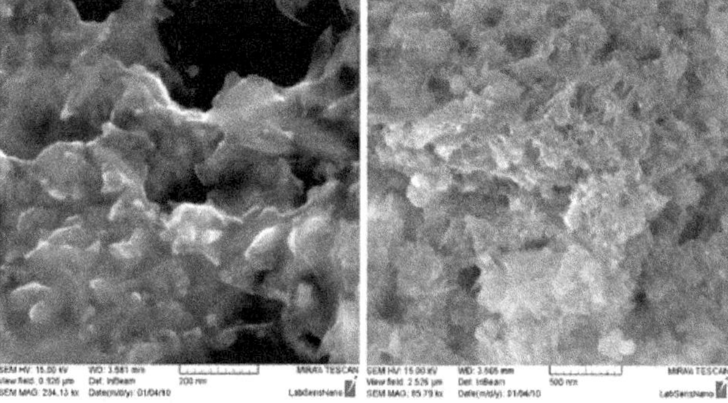

Figure 2: The SEM images of screen-printed polymer paste with Cu_2O/CuO particles before (left) and after (right) firing process.

The electrochemical oxidation of adenine, guanine or other purine derivates at carbon electrode is well known. Formation of complexes of these compounds with metals including copper has been studied. Species Cu(II) can be reduced to Cu(I) and in the presence of adenine, Cu(I) reacts with adenine to form insoluble compounds that accumulate on the electrode surface (Trnkova et al. 2008) and cause decreasing of current response (Fig 3 and Figure 4).

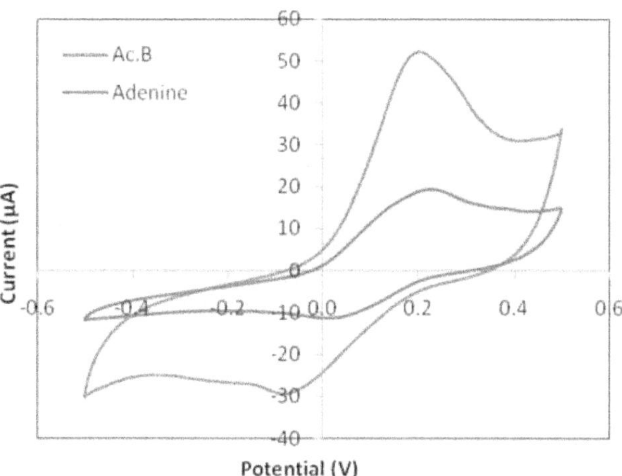

Figure 3: Cyclic voltammogram of cermet electrode with Cu_2O/CuO particles after firing process.

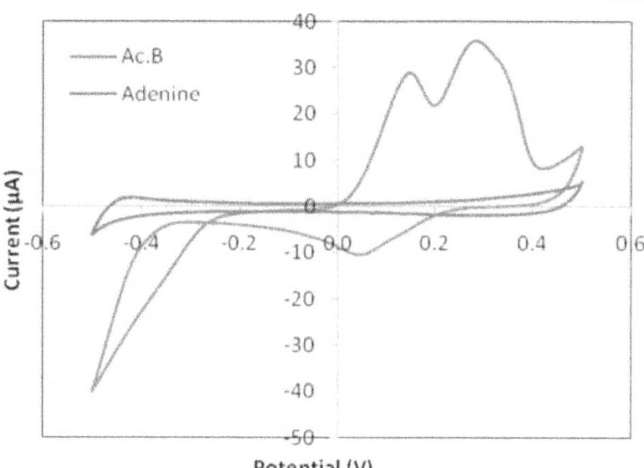

Figure 4: Cyclic voltammogram of cermet electrode with Cu_2O/CuO particles before firing process

The polymer electrode gives bad response due to the polymer binding material that covers Cu_2O/CuO nanoparticles which lead to electrode active area decreasing.

Conclusion

Cu_2O/CuO nanoparticles for preparation of thick film pastes were prepared. Prepared pastes were screen–printed on previously prepared electrode substrate. These electrodes were successfully used as the working electrodes for electrochemical detection of adenine.

Acknowledgement

The financial support from the grant KAN 208130801 and the frame of Research Plan MSM 0021630503 is highly acknowledged.

References

Huang L, Peng F, Yu H, Wang HJ (2009) Preparation of cuprous oxides with different sizes and their behaviors of adsorption, visible-light driven photocatalysis and photocorrosion. Solid State Sciences 11:129-138

Luo YS, Yu BH, Tu YC, Liang Y, Zhang YG, et al. (2008) Self-assembly synthesis of cactuslike Cu2O 3D nanoarchitectures via a low-temperature solution approach. Mater. Res. Bull. 43:2166-2171

Pravda M, O'Meara C, Guilbault GG (2001) Polishing of screen-printed electrodes improves IgG adsorption. Talanta 54:887-892

Trnkova L, Zerzankova L, Dycka F, Mikelova R, Jelen F (2008) Study of copper and purine-copper complexes on modified carbon electrodes by cyclic and elimination voltammetry. Sensors 8: 429-444

Zahmakiran M, Ozkar S, Kodaira T, Shiomi T (2009) A novel, simple, organic free preparation and characterization of water dispersible photoluminescent Cu2O nanocubes. Mater. Lett. 63: 400-402

J Biochem Tech (2010) 2(5):S70-S73
ISSN: 0974-2328

Nanostructured silver and platinum modified carbon fiber microelectrodes coated with nafion for H_2O_2 determination

Vladimir Halouzka, Petr Jakubec, Cenek Gregor, Dalibor Jancik, Gabriela Valaskova, Kyriakos Papadopoulos, Theodor Triantis, Jan Hrbac*

Received: 25 October 2010 / Received in revised form: 13 August 2011, Accepted: 25 August 2011, Published: 25 October 2011
© Sevas Educational Society 2011

Abstract

Carbon fiber microelectrodes equipped with nanostructured metals (platinum and silver) and covered with a Nafion layer constitute sensitive H_2O_2 sensors. Metallic layers on carbon fibers were prepared by surfactant assisted electrodeposition. In the case of silver, the procedure leads to coating which is composed of porous, partially aggregated and crystalline deposits containing silver nanoparticles. The electrodeposition of platinum leads to carbon fiber decorated with clusters of platinum nanoparticles. After coating the electrodes with protective and antiinterference barrier made of Nafion, the sensing properties of the prepared microelectrodes towards hydrogen peroxide are investigated.

Keywords: Carbon fiber, microelectrode, platinum nanoparticles, silver nanoparticles, Nafion, amperometric sensor

Introduction

Carbon fiber electrodes represent an advantageous platform for amperometric sensors fabrication, since they combine microelectrode properties (i.e. enhanced diffusional transport of the analyte onto the electrode surface) with relatively high analytical current signals. Their disadvantage is, however, a slow charge transfer kinetics for some analytes requiring the modification of these electrodes.

Vladimir Halouzka, Petr Jakubec, Gabriela Valaskova, Jan Hrbac*

Department of Physical Chemistry, Palacky University, tr. 17. listopadu 12, 771 46 Olomouc, Czech Republic

Tel: +420 585 634 755, Email: hrbac@aix.upol.cz

Cenek Gregor, Dalibor Jancik

Center for Nanomaterial Research, Palacky University, Slechtitelu 11, 783 71 Olomouc, Czech Republic

Kyriakos Papadopoulos, Theodor Triantis

Institute of Physical Chemistry, NCSR Demokritos, 15310 Agia Paraskevi Attikis, Athens, Greece

Electrodes containing platinum as an active material are frequently used to determine hydrogen peroxide in anodic region of potentials due to the relatively rapid electrode kinetics of hydrogen peroxide electrooxidation on platinum oxides (Hall et al.1998). The mechanism of hydrogen peroxide oxidation on platinum involves the adsorption of intermediate and therefore the dependence of limiting current (for the experiment with rotating platinum electrode) has the character of adsorption isotherm (Hall et al.1998). For this reason, the calibration curve for hydrogen peroxide on smooth platinum electrode deviates from linearity at H_2O_2 concentration as low as 5 mmol·dm^{-3}. Nanostructuring of platinum enables to overcome this problem. A one approach to achieve nanostructuring of the electrode surface is the electrodeposition from the mixture of target metal salt and a suitable structure directing agent. This technique was first described by Evans et al (2002) who prepared platinized platinum electrode from the aqueous solution containing K_2PtCl_6 and octaethyleneglycol monohexadecyl ether and shown that the electrode enables determination of H_2O_2 over a wide concentration range upto 0.1 M.

Nanostructured silver electrodes have recently gained a great deal of interest as sensors for H_2O_2 amperometric determination in the cathodic region of potentials (e.g. Welch et al. 2005; Guascito et al. 2008). Cathodic regime is convenient especially for hydrogen peroxide determinations in biological matrices, since this potential region is usually free from interferences caused mainly by ascorbate and urate, often present in high quantities in biological samples, e.g. body fluids, tissue homogenates, cell lysates etc Kohen et al (2000); Hrbac et al (2000). In this contribution we used polyol – based nonionic surfactants Triton-X100 and Pluronic F127 as structure directing agents for coating carbon fiber microelectrodes with platinum and/or silver. We have found that Pluronic F127 is an effective structure directing agent for coating carbon fiber microelectrodes with silver. On the other hand, Triton-X100 gives optimum results for platinum coatings. In this contribution advantageous properties of silver and platinum material for H_2O_2 sensing are combined with advantages of carbon fiber microelectrodes, i.e. enhanced mass transfer by radial diffusion and relatively large surface area resulting in sufficiently high currents to be monitored using conventional potentiostats.

Materials and methods

For electrochemical measurements the CH Instruments 660C workstation was used in a three-electrode circuit with Pt wire as an auxilliary, Ag/AgCl as a reference and microelectrode sensor as a working electrode. Amperometry in stirred solution was used to test the microelectrode sensors. H_2O_2 aliquots were introduced into the cell using an autosampler. For testing the selectivities of the sensor, aliquots of selected interference compounds were introduced into the cell using Hamilton microsyringes.

Results and discussion

In our efforts to find optimum coating method, we tested a range of surfactants including Triton X100 and Pluronic F127 as additives into platinum and silver plating solutions. The electrodeposition of silver onto carbon fiber was performed from 0.08 M silver nitrate solution containing 25 % of Pluronic F127 at –300 mV vs. Ag / AgCl for 60 sec in quiescent solution. The electrodeposition of platinum onto carbon fiber was performed from 0.01 M K_2PtCl_6 solution containing 25 % of Triton X 100 at –200 mV vs. Ag / AgCl for 60 sec in quiescent solution. The amperograms recorded during the electrodeposition processes are shown in Fig 1 a,b. Fig 2 a,b and 3 a,b show the morphologies of metals electrodeposited onto carbon fibers. In the case of platinum the fiber is decorated with nanometer-sized platinum structures. Similar procedure for silver leads to coverage of the whole of carbon fiber with micron-sized silver crystallites. A closer look at the silver layer reveals that silver nanoparticles can be found attached to the crystallites.

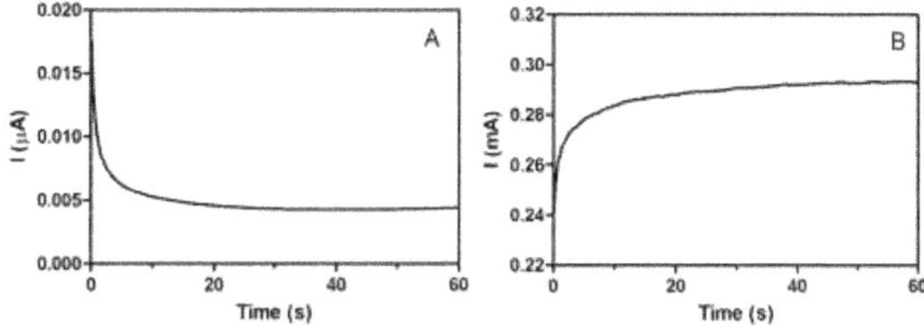

Figure 1: amperograms recorded during electrodeposition processes. A: platinum, deposition process proceeded at -200 mV vs. Ag/AgCl from the solution of K_2PtCl_6 (0.01 M) in the presence of Triton X100 (25 % (w/w)). B: silver, deposition from $AgNO_3$ (0.08 M) in the presence of Pluronic F127 (25 % (w/w))

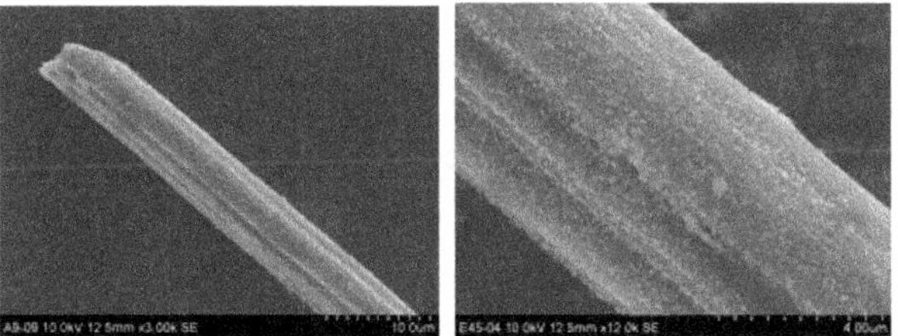

Figure 2: A,B SEM images of carbon fiber decorated with platinum particles

Figure 3: A,B SEM images of carbon fiber covered with silver layer

Coated carbon fibers were tested on their H_2O_2 sensing properties using amperometry at +700 mV vs. Ag/AgCl (platinum covered electrode) and –400 mV vs. Ag/AgCl (silver coated electrode). The resulting performances for electrodes prepared using deposition from solution of target metal salt without structure directing agent (A), Pluronic F127 (B) and Triton X100 (C) are shown in Fig 4 & Fig 5.

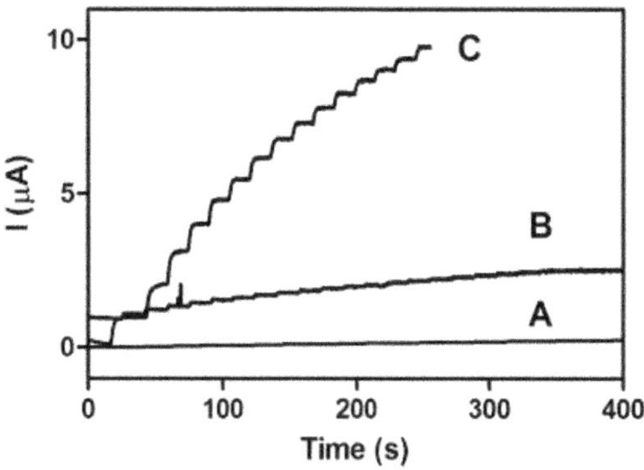

Figure 4: Amperometric curves of hydrogen peroxide for platinum coated carbon fiber electrode prepared prepared using deposition from solution of K_2PtCl_6 without structure directing agent (A), in the presence of Pluronic F127 (B) and Triton X100 (C). Every current step corresponds to the addition of H_2O_2 giving 1 mM increase in H_2O_2 concentration. The measurements were carried out in Britton-Robinson buffer (pH 7) at 700 mV vs. Ag/AgCl.

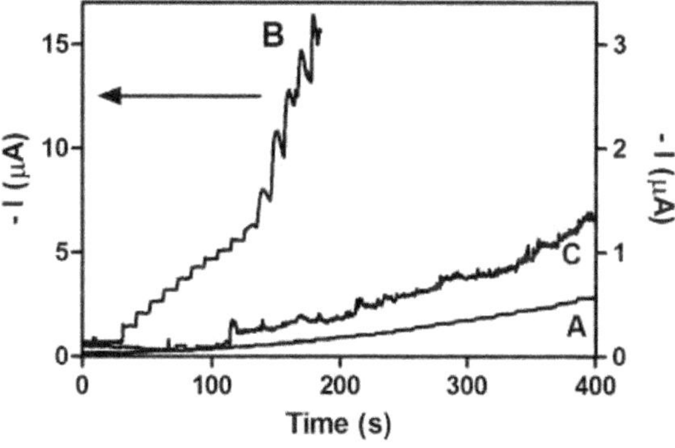

Figure 5: Amperometric curves of hydrogen peroxide for silver coated carbon fiber electrode prepared prepared using deposition from solution of $AgNO_3$ without structure directing agent (A), Pluronic F127 (B) and Triton X100 (C). Every current step corresponds to the addition of H_2O_2 giving 1 mM increase in H_2O_2 concentration. The measurements were carried out in Britton-Robinson buffer (pH 7) at -400 mV vs. Ag/AgCl.

The performance of silver coated electrode depends on chosen operational potential. Amperometric curves for H_2O_2 reduction are shown in Fig 6 for –400 and 0 mV. When the electrode is biased at 0 mV, broader usable concentration range is achieved, at the expense of some current sensitivity.

Long-term (tens of minutes) stabilities of H_2O_2 amperometric responses of both electrode types are rather poor and can be significantly improved by adding protective Nafion layer using a "dip-dry" method (the electrodes were dipped into solution of Nafion and then dried at 80°C for 1h). In Fig 7 the response stability of platinum coated electrode is shown, the corresponding experiment for silver coated electrode is shown in Fig 8.

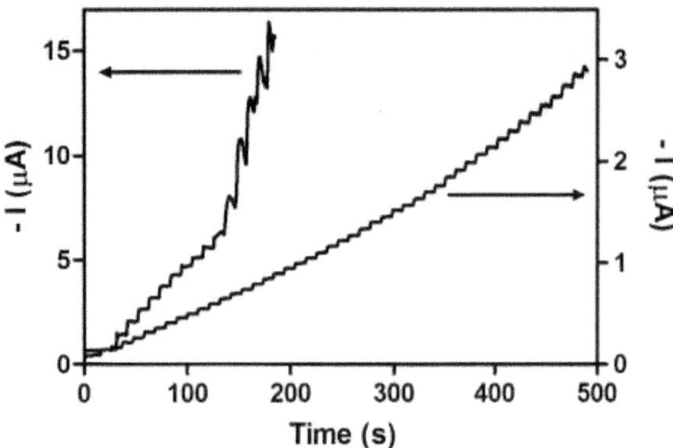

Figure 6: Amperometric response curves of carbon fiber microelectrodes covered with silver and Nafion at - 400 mV (left axis) and 0 mV (right axis) in stirred BR buffer (PH=7), each addition corresponded to 1 mM H_2O_2.

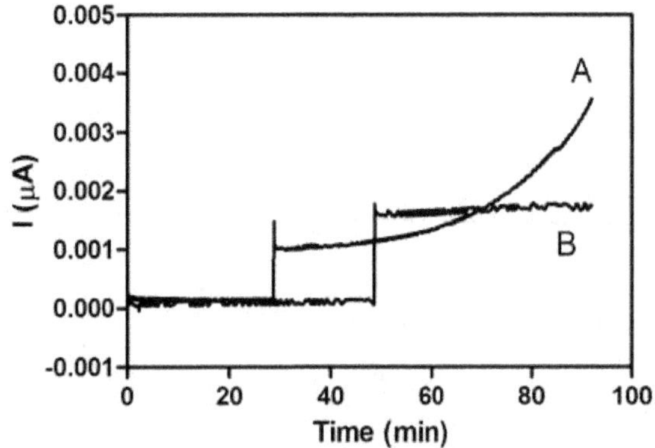

Figure 7: Amperometric response to 20 mM of hydrogen peroxide. A: platinized carbon fiber. B: Electrode prepared the same way equipped with protective layer of Nafion. The measurement was carried out in stirred Britton-Robinson buffer (pH 7) at 700 mV vs. Ag/AgCl.

Carbon fiber microelectrodes covered with silver and Nafion were tested for interferences from easily oxidizable species. Due to the protective layer of Nafion possessing well-known anion-repelling property, ascorbic acid and uric acid do not produce significant interferences. Fig 9,10. The selectivity towards paracetamol (acetaminophen) is lower for platinum-Nafion covered microelectrode.

Conclusion

We have shown that carbon fibers coated with platinum and/or silver and subsequently stabilized by Nafion layer can be used as sensors for hydrogen peroxide amperometric monitoring. The electrodeposition of metallic layers from solutions of platinum and silver salts proved itself to be a suitable coating method. Long-term stable and interference-free operation, as well as a broad range of concentrations, within which the response is linear was achived for both platinum and silver coated carbon fiber microelectrodes.

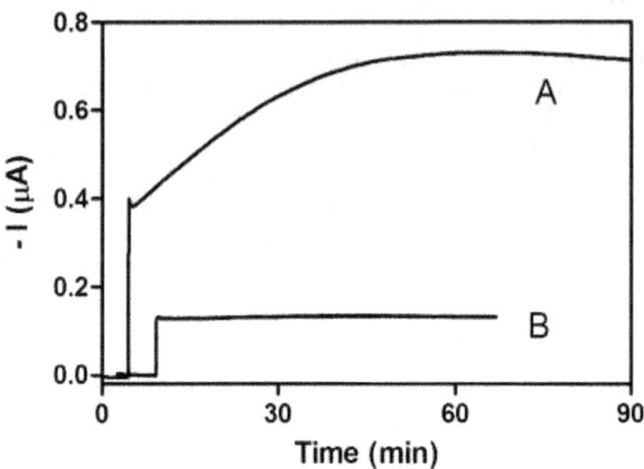

Figure 8: Amperometric response to 20 mM of hydrogen peroxide. A: silver coated carbon fiber. B: Electrode prepared the same way equipped with the protective layer of Nafion. The measurement was carried out in stirred Britton-Robinson buffer (pH 7) at 0 mV vs. Ag/AgCl.

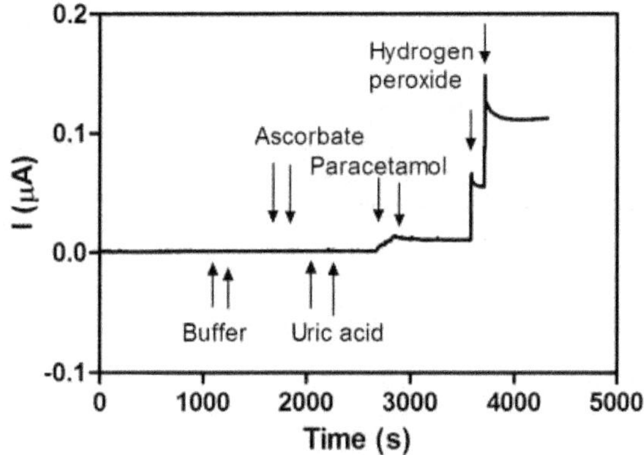

Figure 9: The response of platinum-Nafion coated microfiber to the additions of ascorbate, uric acid, acetaminophen and hydrogen peroxide (each addition indicated by arrow corresponded to 1.00 mM final concentration of each compound). BR buffer, pH=7.0, applied potential: 700 mV vs. Ag/AgCl.

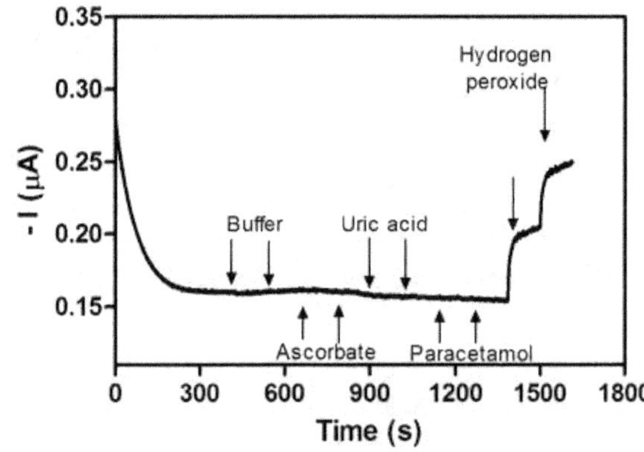

Figure 10: The response of silver-nafion coated microfiber to the additions of ascorbate, uric acid, acetaminophen and hydrogen peroxide (each addition indicated by arrow corresponded to 1.00 mM final concentration of each compound). BR buffer, pH=7.0, applied potential: 0 mV vs. Ag/AgCl.

Acknowledgement

Financial support from the Grant Agency of the Czech Republic (project no. 524/05/P135) and project no. PrF-2010-025-Nanosystemy are gratefully acknowledged.

References

Evans SAG, Elliott JM, Andrews LM et al (2002) Detection of hydrogen peroxide at mesoporous platinum microelectrodes. Anal Chem 74:1322–1326

Guascito MR, Filippo E, Malitesta C et al (2008) A new amperometric nanostructured sensor for the analytical determination of hydrogen peroxide. Biosens Bioelectron 24:1057–1063

Hall SB, Khudaish EA, Hart AL (1998) Electrochemical oxidation of hydrogen peroxide at platinum electrodes. Part 1. An adsorption-controlled mechanism. Electrochim Acta 43:579-588

Hall SB, Khudaish EA, Hart AL (1998) Electrochemical oxidation of hydrogen peroxide at platinum electrodes. Part II: effect of potential. Electrochim Acta 43:2015-2024

Hrbac J, Kohen R (2000) Biological redox activity: Its importance, methods for its quantification and implication for health and disease. Drug Dev Res 50:516-527

Kohen R, Vellaichamy E, Hrbac J et al (2000) Quantification of the overall reactive oxygen species scavenging capacity of biological fluids and tissues. Free Radical Biol Med 28:871-879

Welch CM, Banks CE, Simm AO et al (2005) Silver nanoparticle assemblies supported on glassy-carbon electrodes for the electro-analytical detection of hydrogen peroxide. Anal Bioanal Chem 382:12–21

J Biochem Tech (2010) 2(5):S74-S75
ISSN: 0974-2328

Determination of viral nucleic acids by electrochemical detection array using paramagnetic nanoparticles

Dalibor Huska, Marketa Ryvolova, Jana Chomoucka, Jana Drbohlavova, Vojtech Adam, Libuse Trnkova, Ivo Provazn ik, Jaromir Hubalek, Rene Kizek*

Received: 25 October 2010 / Received in revised form: 13 August 2011, Accepted: 25 August 2011, Published: 25 October 2011
© Sevas Educational Society 2011

Abstract

Simple and express determination of dangerous viruses is one of the most important precautions of their pandemic spreading. Paramegnetic nanoparticles in combination with electrochemical detection methods belong to the group of techniques fulfilling these requirements. We were able to isolate and subsequently detect 100 pg/µl of viral nucleic acid by using 1 µl of paramagnetic nanoparticles.

Keywords: Viral DNA, Electrochemical, Nanoparticles, Nanobiosensor,

Introduction

Development of the simple and fast detection devices like nanobiosensors is one of the main goals in the area of nanotechnology. Application of this technology can be beneficial for effective identification of dangerous viruses such as HIV, influenza and hepatitis. This work is focused on the detection of the hepatitis B virus, which attacks liver and may lead to the cancer. Epidemics all around the world caused that up to one third of the world's population has been exposed to this disease and about 3-6% is currently infected. Magnetic particles responding to an external magnet field are providing an elegant method of separation of targeted molecule from the solution (Huska et al. 2009a, 2009b, 2008).

Taking advantage of their magnetic characteristic and low cost of synthesis, magnetic particles (MPs) have been widely used as a universal separation tool for biologically active compounds such as nucleic acids (i.e., DNA and RNA), proteins and peptides, as well as intact cells. Streptavidin modified paramagnetic particles in conjunction with electrochemical detection have the potential to create highly sensitive and selective detection device

Materials and Methods

Electrochemical DNA analysis was carried out by AUTOLAB analyzer (EcoChemie, Netherlands) in connection with VA-Stand 663 (Metrohm, Switzerland). Viral nucleotides were purchased from Sigma – Aldrich (U.S.A.). Square wave voltammetric (SWV) measurements were carried out in the presence of acetate buffer pH 5.0. SWV parameters: potential step 5 mV, frequency 260 Hz. The analyzed samples were deoxygenated prior to measurements by purging with argon (99.999%), saturated with water for 120 s.

Isolation of nucleic acids proceeded on paramagnetic nanoparticles (MNPs) from Department of Microelectronics, Faculty of Electrical Engineering and Communication, Brno University of Technology. MNPs were biotinilated by viral oligonucleotides (ODNs) and subsequently hybridization with complementary sequences took place. This step was followed by the separation using external magnetic field and releasing of the complementary sequences from the MNPs by thermal deneturation. Isolated oligonucleotides were analyzed by square-wave voltammetry using hanging mercury drop electrode and by chip based capillary gel electrophoresis (Biorad Experion, USA).

Ivo Provaznik

Department of Biomedical Engineering, Faculty of Electrical Engineering and Communication, Brno University of Technology, Kolejni 4, CZ-612 00 Brno, Czech Republic

Dalibor Húska, Markéta Ryvolová, Vojtěch Adam, René Kizek*

Department of Chemistry and Biochemistry, Mendel University in Brno, Zemedelska 1, CZ-613 00 Brno, Czech Republic

*Tel: +420 545 133 350, Fax: +420 545 212 044
E-mail: kizek@sci.muni.cz

Libuše Trnková

Department of Chemistry, Masaryk University, Kotlarska 2, CZ-611 37 Brno, Czech Republic

Jaromír Hubálek, Jana Chomoucká, Jana Drbohlavová

Department of Microelectronics, Faculty of Electrical Engineering and Communication, Brno University of Technology, Udolni 53, CZ-602 00 Brno, Czech Republic

Results and discussion

Suggested method is based on the biotin – streptavidin interaction between biotinilated ODNs used as probes for viral DNA and MNPs modified by streptavidin. Conditions of the isolation were optimized. The optimal hybridization time of the viral nucleic acid to MNPs - 5 minutes - was determined. Eight different sequences of the hepatitis B (responsible for various molecular-biological processes) were analyzed. The dependence of the amount of isolated nucleic acid on the amount of MNPs was studied. For isolation of 100 pg/μl of viral DNA 1 μl MNPs was used. Maximal amount of MNPs was 20 μl, which was able to isolate 2 μg/ml of viral nucleic acid. Both capacities were enough for subsequent electrochemical detection. Isolated viral nucleic acids were detected by electrochemical detection. Linear calibration curves were constructed ($R^2 = 0.99$).

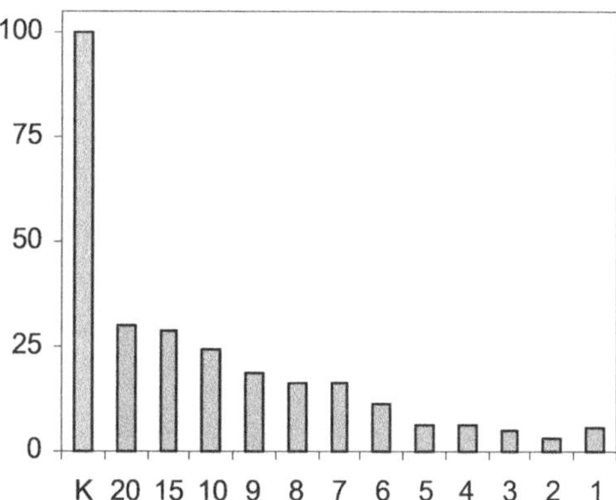

Figure 1: Effect of the amount of MNPs to the total extraction of the viral nucleic acid – electrochemical analysis. (K) bar represents the signal of the control viral DNA sample (5 μg/ml).

Chip-based capillary gel electrophoresis was used for qualitative verification. Finally, experimental electrochemical array was constructed and electrochemical detection as well as electrophoretic analysis was of the isolated viral nucleic acids mixture was performed.

Conclusion

An early identification of the rapidly spreading viruses is one of the preventive actions to protect the world population. Nanobiosensors employing paramagnetic nanoparticles in combination with electrochemical detection methods may be potentially effective way to prevent pandemic spreading of these diseases

Acknowledgements

The work has been supported by NANOSEMED GA AV KAN208130801, NANIMEL 102/08/1546 and GA ČR 102/10/P618

References

Huska D, Adam V, Trnkova L, Kizek R (2009a) The dependence of adenine isolation efficiency on the chain length evidenced using paramagnetic particles and voltammetry measurements. J Magn Magn Mater 321(10):1474-1477

Huska D, Hubalek J et al (2009b) Automated nucleic acids isolation using paramagnetic particles. Talanta 79(2):402 - 411

Huska D, Hubalek J, Adam V, Kizek R (2008) Miniaturized electrochemical detector as a tool for detection of DNA amplified by PCR. Electrophoresis 29(24):4964-4971

J Biochem Tech (2010) 2(5):S76-S78
ISSN: 0974-2328

Heavy metals detection using nanostructured screen printed electrodes

Jan Prasek*, Jana Chomoucka, Jiri Policky

Received: 25 October 2010 / Received in revised form: 13 August 2011, Accepted: 25 August 2011, Published: 25 October 2011
© Sevas Educational Society 2011

Abstract

This work deals with the problematic of heavy metals dissolved in aqueous solution detection using standard electrochemical methods. The problematic of low signal response of miniaturized electrochemical electrodes for electrochemical analysis is mentioned here. Finally some solutions for electrodes miniaturization and examples of fabricated electrodes and their heavy metals detection capabilities on a real sample are shown and discussed here.

Keywords: Screen-printed electrode, Carbon nanotubes, Heavy metals analysis

Introduction

Recently heavy metals analysis is one of the most discussed problems (Ouyang, et al. 2011; Kokkinos et al. 2011; Krystofova et al. 2010). Formerly, since prof. Heyrovsky discovered polarography (1922), heavy metals were detected electrochemically using polarography as a great tool for cheap and fast trace detection in sub ppm and later sub ppb concentrations. From a nowadays point of view, this method in advanced versions is still one of the most reliable and accurate for wide field of usage. The problem is that this method uses toxic mercury as a working electrode material which is in the list of dangerous chemical elements that cannot be used for production in industry. Therefore there is an effort to substitute toxic mercury drop electrode by solid electrodes (Danhel et al. 2011; Atta et al. 2011).

The disadvantage of standard solid electrodes is their robustness which predestines them to be used in laboratories only. The necessity of fast on field or in–vivo detection pushed the electrochemistry to the scene of miniaturized electrochemical systems which usually use planar solid electrode systems. The electrode area of such electrochemical systems usually called as

Jan Prasek*, Jana Chomoucka, Jiri Policky

Dept. of Microelectronics, Brno University of Technology, Udolni 53, 60200 Brno, Czech Republic

*Tel: +420 541 146 192, Fax: +420 541 146 298
E-mail: prasek@feec.vutbr.cz

sensors is in the range of tens of square millimetres (eg. Teng, et al. 2010) The current response of such systems is usually smaller in comparison with standard electrochemical systems due to the active electrode area size of working electrodes which is according to Cottrel equation equal to current response of such systems. In many cases the response of such small electrochemical system is sufficient, but in the case of next miniaturization of electrodes, the signals from the system begin to be insufficient for proper evaluation.

This paper tries to show some possibilities how to solve this problem. Finally an example of heavy metals analysis using modified miniaturized working electrode on a real biological sample is shown here too.

Solutions

One possibility how to solve the problem with the low signal from the miniaturized electrochemical systems mentioned in the introduction is a modification of working electrode. There could be modified material of a working electrode with some biological species that can easily select detected species from the solution. Another possibility is its active area increasing preserving the original geometrical size of the working electrode. This could be achieved using some techniques for 3D structuring of the electrode that could increase the electrode active area several times (Hu et al. 2010; Kalimuthu et al. 2010).

Lithography represents one of the possibilities for nanostructuring process that is usually used for microelectronic devices, but cheaper methods for electrode surface nanostructuring exist. For example vertically aligned direct grown carbon nanotubes on the electrode surface using CVD techniques (Prasek et al. 2006; Wang et al. 2011), aligned nanowires, nanotubes or nanorods created by anodization process (Klosova et al. 2007; Ismail et al. 2011), etc. could be created on the surface of miniaturized electrodes.

One of the most commonly used techniques of nanostructured electrodes preparation is use of nanocomposition made of nanopowder and suitable vehicle. Such nanocomposition could be deposited using screen–printing, dip–coating, drop–coating, spray–casting and other techniques.

Nanostructured electrode example

Example of screen–printed working electrode that was made using a nanocomposition mixture of multi–walled carbon nanotubes (MWNTs) with PMMA as a vehicle is shown in the Fig. 1. The diameter of the electrode is 0.7 mm. The SEM image of fabricated electrode is shown in the Fig 2. From the Fig 2 is clear that the surface of the electrode is covered with high amount of MWNTs which are partially hidden under binding polymer vehicle. This problem is caused by the binding material of the electrode, which probably reduce the electrode active area leading to output current response suppression.

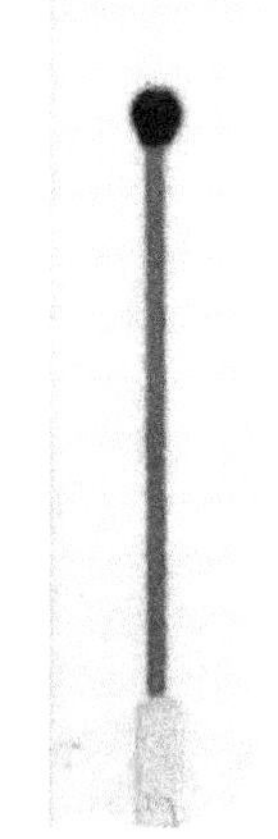

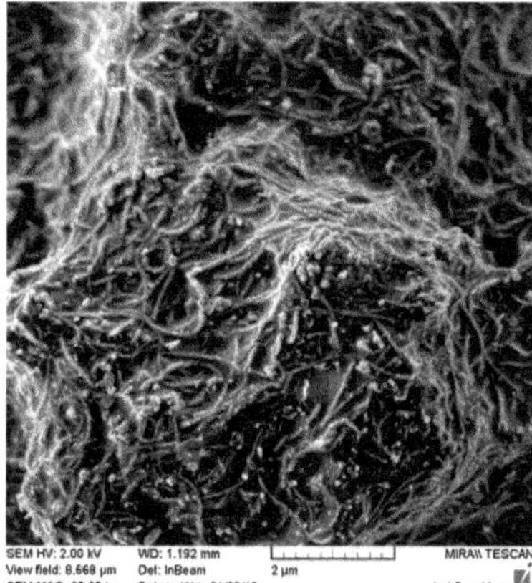

Figure 1: Example of screen–printed electrode with nanostructured surface

Figure 2: SEM image of the electrode surface structure

Heavy metals analysis example

Lead ions detection using the MWNTs based working electrode in a three–electrode system employing differential pulse voltammetry (DPV) is shown in the figure 3. The measured sample was a real sample of the four lead painting differently poisoned aasvogel's eggs extract (from 100 to 1000 µmol/L). The results shown in the Fig. 3 represents voltammetric responses of the electrode to four different concentrations of lead poisoning of aasvogel's eggs. It could be concluded that MWNTs nanocomposition based working electrode is able to detect lead ions on real sample satisfactorily from 100 µmol/L.

Although the obtained results are satisfactory for this application it was expected that the detection limit could be better. Therefore other experiments of the laboratory sample solution containing lead were done. It was found that the detection limit of this electrode is in the scale of units of µmol/L. This result is very good considering the surface of the electrode covered with polymer binding material which probably causes reduction of electrode active area. The result is also comparable to other standard electrodes. It could be concluded that after little optimization of the binding material, this electrode could be used as a working electrode of electrochemical voltammetric sensors for detection of species dissolved in aqueous solutions.

Conclusion

The possibilities of standard electrodes miniaturization into the small hand held systems using nanostructured microelectrodes were shown and discussed in this paper. An example of the screen–printed electrode, its surface structure and the current DPV response to real biological sample of the lead painting poisoned aasvogel's

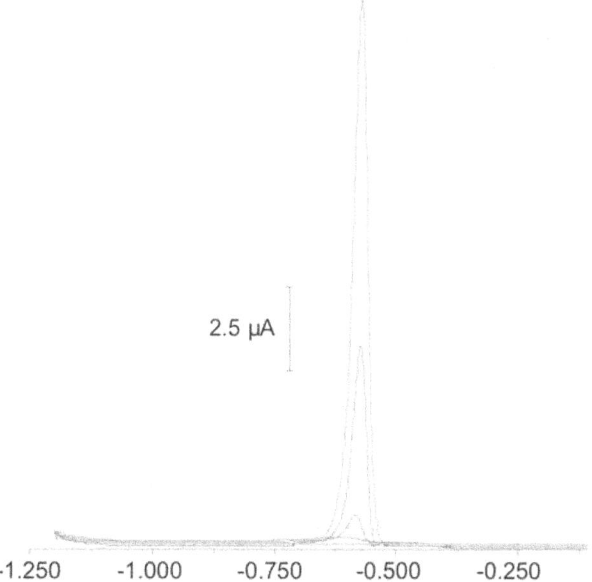

Figure 3: Example of electrode with nanostructured surface response to lead contained in the lead painting poisoned aasvogel's eggs extract.

eggs extract with satisfactorily results is shown here too. Using the standard laboratory sample, it was found that the detection limit of fabricated electrode is in the scale of units of µmol/L which is comparable to the results obtained with standard solid electrodes. This confirms our presumption about the suitability of MWNTs based screen-printable nanocomposition to be used as the working electrode of electrochemical sensors.

Acknowledgement

The work has been supported by Czech grant project GACR 102/09/P640 and the research plan MŠM 0021630503 MIKROSYN.

References

Atta NF, Galal A, Ahmed RA (2011) Direct and Simple Electrochemical Determination of Morphine at PEDOT Modified Pt Electrode. Electroanalysis 23(3):737-746

Danhel A, Yosypchuk B, Vyskocil V, et al. (2011) A novel paste electrode based on a silver solid amalgam and an organic pasting liquid. Journal of Electroanalytical Chemistry 656(1-2):218-222

Hu Y, Huang X, Wang K, et al. (2010) Kirkendall-effect-based growth of dendrite-shaped CuO hollow micro/nanostructures for lithium-ion battery anodes. Journal of Solid State Chemistry 183(3): 662–667

Ismail S, Razak KA, Jing PW, et al. (2011) Tungsten Oxide Nanoporous Structure Synthesized Via Direct Electrochemical Anodization. Enabling science and Nanotechnology 1341: 21-24

Kalimuthu P, John SA (2010) Simultaneous determination of ascorbic acid, dopamine, uric acid and xanthine using a nanostructured polymer film modified electrode. Talanta 80(5): 1686–1691

Klosova K, Hubalek J (2008) Advanced electrodes with nanostructured surfaces for electrochemical microsensors. Physica Status Solidi A 205(6): 1435–1438

Kokkinos C, Economou A (2011) Disposable Nafion-modified micro-fabricated bismuth-film sensors for voltammetric stripping analysis of trace metals in the presence of surfactants. Talanta 84(3):696-701

Krystofova O, Trnkova L, Adam V, et al. (2010) Electrochemical Microsensors for the Detection of Cadmium(II) and Lead(II) Ions in Plants. Sensors 10(6):5308-5328

Ouyang RZ, Zhu ZQ, Tatum CE, et al (2011) Simultaneous stripping detection of Zn(II), Cd(II) and Pb(II) using a bimetallic Hg-Bi/single-walled carbon nanotubes composite electrode. J Electroanal Chem 656(1-2): 78-84

Prasek J, Hubalek J, Adamek M, et al (2006) Nanopatterned working electrode with carbon nanotubes improving electrochemical sensors. 5th IEEE Sensors Conference 1257–1260

Teng YJ, Chen C, Zhou CX, et al. (2010) Disposable amperometric biosensors based on xanthine oxidase immobilized in the Prussian blue modified screen-printed three-electrode system. Science China-Chemistry 53(12):2581-2586

Wang W, Epur R, Kumta PN (2011) Vertically aligned silicon/carbon nanotube (VASCNT) arrays: Hierarchical anodes for lithium-ion battery. Electrochemistry Communication 13(5): 429-432

J Biochem Tech (2010) 2(5):S79-S80
ISSN: 0974-2328

Applications of nano particles in increasing sensor signals

Eva Svabenska

Received: 25 October 2010 / Received in revised form: 13 August 2011, Accepted: 25 August 2011, Published: 25 October 2011
© Sevas Educational Society 2011

Abstract

Different types of elelectrochemical sensors were studied from the point of view of possibility to enhanced signal response by using gold nanoparticles and carbon nanotubes. The measurement was performed by using amperometry and cyclic voltamometry. For testing, gold and platinum porous or nonporous and graphite sensors were taken. All tested layers with nanoparticles, especially with carbon nanotubes, provided higher response comparing to layers without nanoparticles. Sensors from porous platinum gave the highest response.

Keywords: nanoparticles, sensor, cyclic voltametry, amperometry

Introduction

The aim of our research is to increase response of electrochemical sensors by using nanoparticles. Nanoparticles could cause increase of the working surface of the electrode and improve interactions between the redox active compounds, enzymes and the electrode. In our present experiment, we use different types of sensors with 4 working gold electrodes prepared by screen-printing technology. The sensors were modified by different combinations of peroxidase, albumin, glutaraldehyde and nanoparticles. The mixture was applied on a pure surface of electrode or by means of cystamine–based self–assembled monolayer. Evaluation of the activity of immobilized peroxidase was measured by amperometry and cyclic voltammetry (CV). CV was realized by comparing voltamograms in buffer and the presence of a redox probe (Skládal et al. 2010).

Materials and Methods

Glutaraldehyde (GA, 25% aq. solution), bovine serum albumin (BSA), cystamine, gold nanoparticles (5nm, 20 nm), Single-Walled

Eva Svabenska

VOP–026 Šternberk,s.p.,VTÚO Brno Division, Czech Republic

*Tel: + 549 49 7010, Fax: +420 541 129 506
E-mail: 364177@mail.muni.cz

Carbon Nanotube (SWCN) and hexaammineruthenium (III) chloride were obtained from Sigma. Potassium ferricyanide, potassium iodide, hydrogen peroxide were obtained from Penta. Phosphate buffer (50mM sodium phosphate, pH adjusted to 7.0), phosphate buffered saline (PBS, 50mM sodium phosphate pH 7.0 and 150mM sodium chloride), acetate buffer pH 4 (50mM sodium acetate, pH 4 with 150mM NaCl) were used for most experiments.

Amperometry measurements was realized by prototype four-channel potentiostat (1.25μA range, ±1V excitation potential) ImmunoSMART. This system combines the measuring part including four miniature peristaltic pumps, digital microcontroller and battery. Control computer with software ML-615x is connecting by serial RS232C link. The amperometric response was measured in 0.5mM H_2O_2 mixed with the zone of 1mM potassium iodide; the iodine produced by the peroxidase label was detected on the sensor at -50mV vs. the silver pseudo reference electrode.

Cyclic voltammetry was measurement by commercial device EmStat (Palm Instruments) with six current ranges from 1 nA to 100 uA full scale and with a minimum resolution of 1 pA. Electrochemical measurements were performed in three electrode cell, where a platinum wire was used as the auxiliary electrode and Ag/AgCl electrode as the reference electrode. Our sensors were used as working electrode. 2mM concentration of $K_3[Fe(CN)_6]$ and $Ru(NH_3)_6Cl_3$ in PBS and 1mM concentration H_2O_2 in acetate buffer pH 4 were used as the redox probes. Scan window was between -0.6 to 0.6 V or -0.8 to 0.4 V for H_2O_2, scan rate 50mV/s. For evaluation each of sensors the difference between baseline and measurements with redox probe was taken, while the potential values were - 500mV or – 400mV for H_2O_2.

Results and Discussion

In Table 1, you can see that we changed type of working electrode, type and size nanoparticles during a imobilization procedure of plotting layers. The non specific binding sites were saturated by bovine serum albumin.

The results obtained by CV and amperometric measurement of sensors shown some increase in enzyme activity and porosity of the layers. Au nanoparticles provided increase of the current response, but not in all tested configurations. The use of carbon nanotubes seems to be promising configuration. Sensors with porous Pt layer give higher responses than the other plain-surface sensors.

Table 1: Configuration of individual sets

Set	Working electrode	Nano-particles	Obtained response	
			Amp (nA)	CV (nA)
A	Au	5nm, 20nm Au	870	750
B	Au	20nm Au	559	903
C	Au	5nm, CN	1250	1425
D	Pt – p,n; Au – p,n	–	1250	2295
E	Pt – p,n; Au – p,n	–	31,6	770
F	graphite, Au	5nm, 20nm Au	742	1702

P = porous, n = non–porous; CN = carbon nanotube; Amp. = amperometry

Conclusion

The highest response we got for combination with gold working electrode and carbon nanotubes and for platinum electrode without nanoparticles. The best sensors modifications are going to be used for detection of bacteria by using specific antibodies.

Acknowledgement

The work has been supported by Ministry of Defense of Czech Republic (projects no. OVVTUO2008001).

References

Skládal P, Pohanka M, Kupská E, Šafář B (2010) Biosensors for Detection of Francisella tularensis and Diagnosis of Tularemia. Biosensors Serra PA, Ed, In–Tech pp 302

J Biochem Tech (2010) 2(5):S81-S82
ISSN: 0974-2328

Ta$_2$O$_5$ nano-crystals created by anodization

Marina Vorozhtsova, Radim Hrdý, Jaromír Hubálek*

Received: 25 October 2010 / Received in revised form: 13 August 2011, Accepted: 25 August 2011, Published: 25 October 2011
© Sevas Educational Society 2011

Abstract

Nanostructures of Ta$_2$O$_5$ have a great potential in the next generation semiconductor electronics. These structures can be prepared by anodization of Ta through an Al$_2$O$_3$ template under certain conditions. Nanocrystals created in this experiment were examined by the scanning electron microscopy (SEM). The next experiment will be the creation of Ta$_2$O$_5$ nanowires.

Keywords: Ta$_2$O$_5$ nanocrystals, anodization

Introduction

Ta$_2$O$_5$ is a material of great interest for fabricating semiconductor and photonic devices. This is due to its unique properties such as high dielectric constant, low leakage current density, high index of refraction and low optical propagation losses. Its high dielectric constant and low leakage current density make it popular for a use in the next generation semiconductor electronics (Hrdy et al. 2007).

In certain electrolytes for porous alumina formation the anodic processing (under certain conditions) of a specimen, which consists of a Ta layer covered with a relatively thick Al layer, results in the formation of metal oxide nanocrystals, which are systematically self–organized in the depth of the alumina pores, as outlined in Fig 1.

Materials and methods

The anodization of 2 µm thick aluminum layer, which was sputtered on the tantalum, proceeded at 16 °C in oxalic acid and led to a creation of nanopores with size of about 50 nm. Anodization time was about 13 minutes. During anodization, the steady–state voltage is adjusted to 53 V in order to grow the film with a particular pore size and pore population density.

Marina Vorozhtsova, Radim Hrdý, Jaromír Hubálek*

Department of Microelectronics, Faculty of Electrical Engineering and Communication, Brno University of Technology, 602 00 Údolní 53, Brno

*Tel: +420 541 146 163, Fax: +420 541 146 298
E-mail: hubalek@feec.vutbr.cz

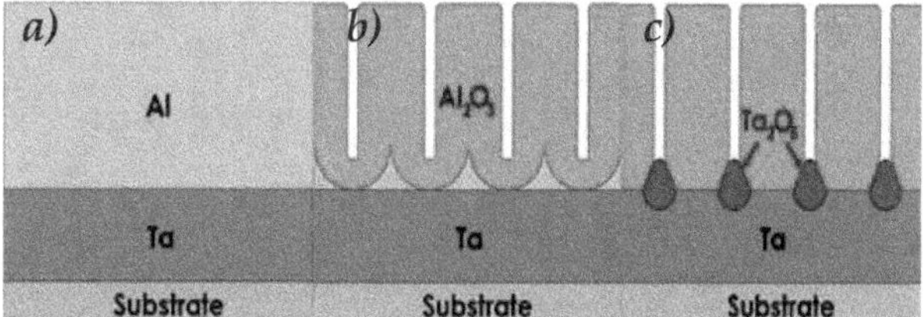

Figure 1: Schematic diagram showing the main steps for forming Ta$_2$O$_5$ nanocrystals a) a sample before anodization b) porous anodizing the Al layer down to the Ta layer c) locally anodizing the Ta layer through the pores (adapted from Mozalev et al., 2009).

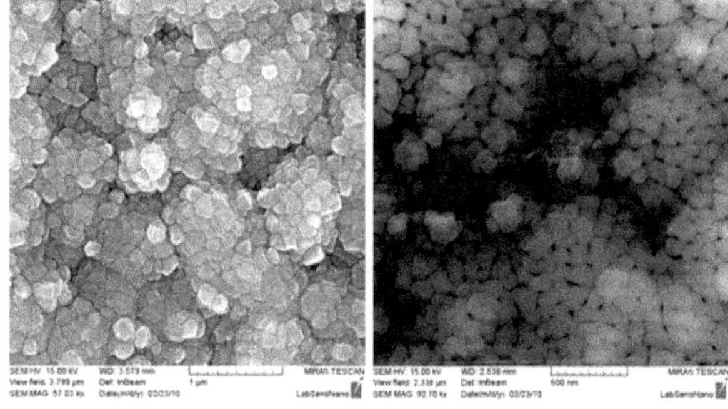

Figure 2: SEM pictures of the sample surface a) layer of aluminum before anodization b) porous alumina surface anodized in oxalic acid at 53 V at 16 °C. Anodization time was 13 minutes. The scale bar is 1µm (left) respectively 500 nm (right).

Results and Discussion

Nanocrystals of the tantalum oxide (hillocks) begin to grow when the alumina barrier layer reaches the tantalum layer (Fig 1b); since then the alumina barrier layer is dissolved at the alumina/tantalum interface by the growing of the tantalum oxide nanocrystals.

The sample was examined by the scanning electron microscopy (SEM) Tescan FE Mira II LMU in high vacuum mode 1.2 x 10^{-2} Pa and voltage 15 kV. A created structure is shown in Fig 3 .

Conclusion

Nanocrystals of tantalum oxide have been created via anodization of an aluminum layer and following dissolution of the porous template. The next step is going to be a continuation of the method mentioned above and the creation of Ta_2O_5 nanowires.

Figure 3: SEM pictures of Ta_2O_5 nanocrystals on crystals of Ta (after dissolution of the alumina template), the scale bar is 1μm (left) respectively 500 nm (right) .

Acknowledgement

This research was supported by the Czech Ministry of Education the frame of Research Plan MSM 0021630503.

References

Hrdý R, Klosová K, Hubálek J (2007) New type of micro sensor with active nanoparticle surface, EMPC, Oulu, Finland, pp. 195–197

Mozalev A, Smith AJ, Borodin S, Plihauka A, Hassel AW et al. (2009) Growth of multioxide planar film with the nanoscale inner structure via anodizing Al/Ta layers on Si. Electrochimica Acta 54:935–945

J Biochem Tech (2010) 2(5):S83-S84
ISSN: 0974-2328

Expressing of bacterial dihydrodipicolinate synthase in transgenic barley

Natalia Cernei, Ondrej Zitka, Ludmila Ohnoutkova, Katarina Mrizova, Petr Galuszka, Jana Vasková, Mark A. Smedle y, Wendy A. Harwood, Petr Babula, Rene Kizek*

Received: 25 October 2010 / Received in revised form: 13 August 2011, Accepted: 25 August 2011, Published: 25 October 2011
© Sevas Educational Society 2011

Abstract

Nutritional quality of human and animal foodstuffs is determined by the content of essential amino acids. Barley is the fourth most important cereal of the world and the second most important cereal grown in the Czech Republic. Cereal grains such as barley contain insufficient levels of some essential amino acid, especially lysine. The ionex chromatography is very selective and convenient for aminoacids determination such as lysine. In connection with post column derivatization the limit of detection of lysine determination was below 1µM and linearity of calibration curve was $R^2=0.995$. Thus we were able to determine the rate of expression for transgenic Barley.

Keywords: Lysine, Barley, DHPS, gene expression

Natalia Cernei, Ondrej Zitka, Rene Kizek*

Department of Chemistry and Biochemistry, Faculty of Agronomy, Mendel University in Brno, Zemedelska 1, CZ-613 00 Brno, Czech Republic

*Tel: +420 545 133 350, E-mail: kizek@sci.muni.cz

Jana Vasková

Institute of Experimental Botany, v.v.i., Academy of Sciences of the Czech Republic

Ludmila Ohnoutkova

Department of Cell Biology and Genetics, Faculty of Science, Palacky University in Olomouc, Slechtitelu 11, 78371 Olomouc-Holice, Czech Republic

Katarina Mrizova, Petr Galuszka, Mark A. Smedley

Department of Biochemistry – Division of Molecular Biology, Faculty of Science, Palacky University in Olomouc, Czech Republic

Wendy A. Harwood

Department of Crop Genetics, John Innes Centre, Norwich Research Park, United Kingdom

Petr Babula

Department of Natural Drugs, Faculty of Pharmacy, University of Veterinary and Pharmaceutical Sciences, Palackeho 1-3, CZ-612 42 Brno, Czech Republic

Introduction

Provision of sufficiently quality food and fodders is still present appeal. World food crisis brings to border of famine millions of people. Food supplementation for biologically important compounds, such as vitamins, amino-acids and essential fatty acids may contribute for solving of this problem. Modern methods of molecular biology enable introducing of new properties into plants. Lysine is biologically very important amino-acid, which is essential for human. Adult needs about 1-1.5 g of lysine per day, children about 44 mg of lysine per day. Generally, L-lysine is essential structural amino-acid, which is crucial for all proteins. Lysine plays important role in calcium uptake absorption, participates in biosynthesis of hormones, enzymes and antibodies and its importance is also in metabolism of muscular tissue.

Positive effect of lysine is demonstrated in processes of healing of wounds after surgical interventions or injuries. Due to these properties, supplementation of fodders for livestock by lysine is very important factor for improving of daily increase in weight (Krishnakumar et al. 2010).

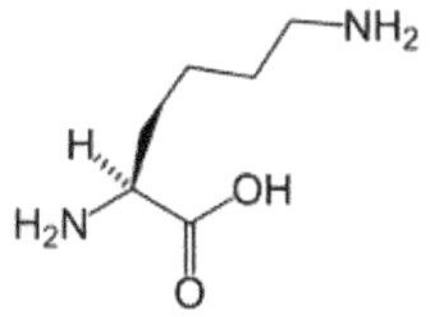

Figure 1: Structure of lysine.

Materials and methods

Transgenic barley plants of the T_0 generation were evaluated by PCR, Real-Time PCR and Western blot. Amino acids content was analyzed by Aminoacid Aminoanalyzer AAA400 with post column derivatization by ninhydrin. Leave samples were prepared by HCl hydrolysis: 0.1g of sample with 2 ml 6M HCl was mineralized for 24 thermoboxes in 85 °C. Glass column with inner diameter 3.7 mm and 350 mm length was filed manually with strong catex ionex in natrium cycle LG ANB with approximately 12 µm particles and 8% porosity. Column was tempered in range 35 - 95°C. Double channel VIS detector with inner cell volume 5 µl worked statically under two wavelengths 440 and 570nm. Solution of ninhydrin (Ingos, Czech Republic) was prepared by dilution in 75 % v/v metylcelosolve (Ingos, Czech Republic) and in 25 % v/v 4M acetic buffer (pH5.5). For reduction $SnCl_2$ (Lachema, Czech Republic) was used. Prepared solution of ninhydrin was stored under inert atmosphere (N_2) in dark and cooled. Elution of Lysine was done by buffer contains natrium citrate 19.6g, NaCl 52.6g and Boric acid 2.05g x $1L^{-1}$ and pH was adjusted to 9.7 by NaOH 0.2M. Flow rate was 0.3 ml/min. Reactor temperature was 120°C.

Results and discussion

Two constructs pBract214::sTPdapA and pBract214::mdapA containing the dapA gene from *Escherichia coli* coding bacterial DHPS were used for transformation of barley. The vector pBract214::sTPdapA in addition includes the transit peptide Rubisco *Hordeum vulgare* ribulose-1,5-bisphosphate carboxylase small subunit, Genbank U43493. An Agrobacterium-mediated technique was used for transformation of immature embryos of barley cv. Golden Promise. Plants were analysed with respect to expression of cloned genes by qRT-PCR technique. On basis of this analysis, four groups of plants were established (low, medium, high and unequal gene expression).

Due to this fact, we decided to carry out analysis of amino-acid lysine as product of gene transcription. Ionex chromatography was used for analysis. Signal of lysine was well-separated with limits of detection about 500 nM and quantification about 5 µM with error determination about 5%. Real samples in 0.025g weight were hydrolyzed by 0.5ml HCl (6M) in microwave under 80W in time 160minutes. Obtained extracts were subsequently applied into chromatographic system in triplicate (R.S.D. was 6.5%). Concentration of lysine in control group of plants was about 4.5% of total content of all amino-acids. From obtained results follows very good positive correlation between rate of expression of cloned gene and lysine concentration ($R2 =0.99$). We observed that content of lysine was the highest by RT-High expression in compare with control which has the lowest content of Lysine. RT-different expression has similar content of lysine to RT-middle expression and RT-low expression was almost in the middle about content of lysine between Control and RT-midle expression (Fig. 2).

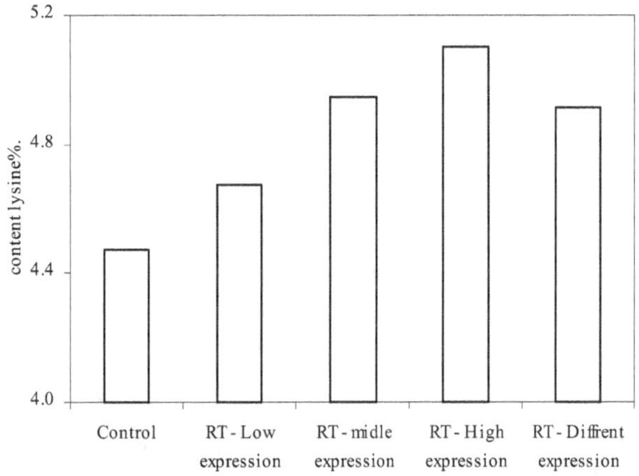

Figure 2: The volume-based lysine (%) on the expression of various genetically modified plants to lysine

Conclusion

Amino acids content was analyzed by Aminoacid Aminoanalyzer AAA400 with post column derivatization by ninhydrin. 30 samples were measured with the gene expression of lysine. The result was converted to % of total lysine. Lysine content had no effect on other amino-acids.

Acknowledgements

This work was supported by project 1M06030 of the Ministry of Education, Youth and Sports of the Czech Republic and REMEDTECH GA ČR 522/07/0692.

References

Krishnakumar V, Manohar S, Nagalakshmi R Krishnakumar et al (2010) Spectrochimica acta Part A Molecular and biomolecular spectroscopy. L-lysine sulphate and characterization. Spectrochimica Acta Part A-Molecular And Biomolecular Spectroscopy 75(5):1394-1397

J Biochem Tech (2010) 2(5):S85-S86
ISSN: 0974-2328

Determination of serum prostate specific antigen in patients with tumours of prostate and tumour cell lines

Natalia Cernei, Michal Masarik, Jaromir Gumulec, Ondrej Zitka, Petr Babula, Rene Kizek*

Received: 25 October 2010 / Received in revised form: 13 August 2011, Accepted: 25 August 2011, Published: 25 October 2011
© Sevas Educational Society 2011

Abstract

Tumour markers are biochemical indicators of malignant tumour proliferation. They serve not only for diagnostics of malignant tumour diseases, but also for monitoring of effect of therapy and recurrence of disease. Serum prostate specific antigen (PSA) belongs to group of the best tumor markers, what are available to medicine at the present. In our study the Immune Enzymatic Automated Analyzer AIA 600 II was employed. Calibration curves for PSA and fPSA demonstrates very good linearity by values of R^2 = 0.999 in both cases. We assumed to determine levels of PSA in real samples of patients and compare to cell lines. In comparism with non-tumour cell lines we observed the enormous growth of PSA 129-times and fPSA 170-times by RVL-22 and negligible growth for PC-3 cell lines about 2 times. Correlation of obtained data (method PSA2 and method PA) was very close with R^2 =0.990.

Keywords: PSA, fPSA, oncological markers, prostate carcinoma.

Introduction

Serum prostate specific antigen (PSA) has been first time used as a screening test for prostate cancer in 1991 by Catalona et al. PSA is

Natalia Cernei, Ondrej Zitka, Rene Kizek*

Department of Chemistry and Biochemistry Mendel University in Brno, Zemedelska 1, CZ-613 00 Brno, Czech Republic

*Tel: +420 545 133 350, Fax: +420 545 212 044
E-mail: kizek@sci.muni.cz

Michal Masarik, Jaromir Gumulec

Department of Pathological Physiology, Faculty of Medicine, Masaryk University, Komenskeho namesti 2, CZ-662 43 Brno, Czech Republic

Petr Babula

Department of Natural Drugs, Faculty of Pharmacy, University of Veterinary and Pharmaceutical Sciences, Palackeho 1-3, CZ-612 42 Brno, Czech Republic

glycoprotein composed of 237 amino-acids, which originates in prostate gland and is necessary to normal physiological function of sperm. At the physiological state, most of PSA is secernated into sperm, because it causes sperm fluidification. Only small amount of PSA is transported into blood, due to this fact, PSA is detectable in blood serum. In the case of disorganization of inner architecture of prostate gland, majority of PSA is transported into blood, which results in increased PSA level in blood serum. It was determined that increased PSA level is associated with tumour diseases of prostate gland. Determination of PSA level is at the present the best prostate tumour marker, which is presently known and used.

Materials and methods

Samples were analyzed by the use of apparatus Immune Enzymatic Automated Analyzer AIA 600 II, which serves for measurement of immunochemical parameters in biological liquids; it uses set of reagents AIA–PACK PSA and fPSA. Blood serums of patients suffering from prostate adenocarcinoma were obtained from St. Anne's University Hospital Brno, all with permission of ethic commission. Cell lines were derived from normal prostate tissue (PNT1A) and tumour tissue - tumour cell lines PC-3 and RV-1. Cells were maintained at 37°C in a humidified incubator with 5% CO_2. The passages of all cell lines were within the range 10-35. Once the cells grew up to 50-60% confluence of the culture and grow medium was replaced by fresh medium for 24 h to synchronize cell growth. In our experiment, we used fully automated immunochemical detection of serum prostate specific antigen. Reaction itself is initiated by pipetting of sample (10 µl) consisting of blood serum or cell lysate with diluents in testing pot AIA-PACK. Samples were incubated at 37°C, antibodies were bonded on surface of paramagnetic particles. Separation of bonded and unbounded antibodies is attained by washing by rinsing solution - unbounded antibodies were washed out. After washing step, substrate - 4-methylumbelliferyl phosphate was added into testing pot and enzymatic activity on paramagnetic particles was fluorescence measured.

Results and discussion

In presented study we compared two methods for determination of total PSA and free fPSA. Calibration curve for PSA was designed

(R^2 = 0.9993, determination error 0.2 %) and fPSA (R^2 = 0.9990, determination error 0.5 %). Firstly we analyzed 26 samples of patients suffering from prostate adenocarcinoma and 3 controls (healthy young men). Levels of total PSA were determined using two procedures PSA2 and PA. Correlation of obtained data (method PSA2 and method PA) was very close with r =0.990. Newly introduced method PSA2 decreased value of determined PSA (method PA) for 1.5 ng/ml at average. In addition, in higher PSA values determined by PA method, PSA level was for about 25-30 % lower in comparison with PSA2 method. At low PSA levels, higher sensitivity of PSA determined by PSA2 method was observed (enhancement of PSA for about 5-7 % in comparison with PA method).

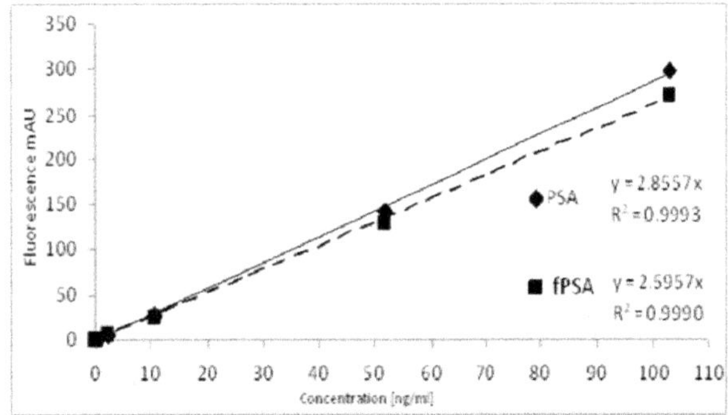

Figure 1: Calibration curve for determination of serum prostate specific antigen of total (PSA) and free (fPSA) PSA.

Proposed methodology for determination of PSA and fPSA was subsequently tested on cell lysates. We determined that proposed methodological procedure is applicable for determination of PSA and fPSA to one million of cells. In non-tumour cell lines, PSA levels were about 0.02 ng/ml and for fPSA 0.02 ng/ml. In the case of tumour cell lines - for tumour cell line PC-3, PSA level was 0.05 ng/ml and level of fPSA was 0.02 ng/ml; tumour cell line RV-1 22 demonstrated significant enhancement of PSA levels to 2.58 ng/ml and fPSA to 3.40 ng/ml.

Conclusion

Determination of serum prostate specific antigen in blood serum of patients is common diagnostic method. We demonstrated in our work that fully automated technique of PSA immunodetection is possible also in cell lysates.

Acknowledgements

We would like to thank previous and current members of the Broach laboratory for helpful discussions for this project. This work was supported by IGA MZ 10200-3 and IGA MENDELU 1/2011.

References

Catalona WJ, Smith DS, Ratliff TL et al (1991) Measurement of prostate-specific antigen in serum as a screening test for prostate cancer. New Eng J Med 324:1156-1161

J Biochem Tech (2010) 2(5):S87-S88
ISSN: 0974-2328

Isolation of DNA by the help of robotized detection on paramagnetic particles

Zdenek Janicek, Dalibor Huska, Libuse Trnkova, Ivo Provaznik, Jaromir Hubalek, Rene Kizek*

Received: 25 October 2010 / Received in revised form: 13 August 2011, Accepted: 25 August 2011, Published: 25 October 2011
© Sevas Educational Society 2011

Abstract

High-quality DNA isolation is the most important step in its study. This step is without questions enabled by paramagnetic particles (MPs). Paramagnetic particles bring to the traditional DNA isolation procedures many new possibilities, above all automation of the complete process. In this work, we were focused on optimization of DNA isolation by the help of paramagnetic particles with use of automated pipetting station.

Keywords: Paramagnetic particles, DNA isolation, Electrochemical detection,

Introduction

Automation enables especially significant reduction of contamination of sample and experimental mistakes caused by

Zdenek Janicek, Ivo Provaznik

Department of Biomedical Engineering, Faculty of Electrical Engineering and Communication, Brno University of Technology, Kolejni 4, CZ-612 00 Brno, Czech Republic

Dalibor Huska Rene Kizek*

Department of Chemistry and Biochemistry, Mendel University in Brno, Zemedelska 1, CZ-613 00 Brno, Czech Republic

*Tel: +420 545 133 350, Fax: +420 545 212 044
 E-mail: kizek@sci.muni.cz

Libuse Trnkova

Department of Chemistry, Masaryk University, Kotlarska 2, CZ-611 37 Brno, Czech Republic

Jaromir Hubalek

Department of Microelectronics, Faculty of Electrical Engineering and Communication, Brno University of Technology, Udolni 53, CZ-602 00 Brno, Czech Republic

human factor. Proposition of this procedure is very important for both diagnostic and criminalistic purposes (Huska et al. 2008, Huska et al. 2009). Due to this fact, sample collection and subsequent isolation of analyte represent one of the crucial moments of supposed analysis. In consideration of increasing pressure on reduction of quantity of material available for analysis (hair, drop of blood serum or full blood, several cells of tissue), possibility of contamination of sample by personnel increases (Huska et al. 2009). Obtained false positive result subsequently can influence findings of investigation or proposal of therapeutic procedure. Possible change into process of nucleic acids isolation can bring paramagnetic microparticles (Palecek and Fojta 2007). As it is has been demonstrated, connection of isolation of nucleic acids by paramagnetic particles with subsequent electrochemical detection brings simple technological procedure (Palecek et al. 2002). Aim of this work consisted in testing of differently modified paramagnetic microparticles for DNA isolation.

Materials and methods

Electrochemical DNA analysis was carried out by the help of apparatus AUTOLAB analyzer (EcoChemie, Netherlands) in connection with VA-Stand 663 (Metrohm, Switzerland). Square wave voltammetric (SWV) measurements were carried out in the presence of acetate buffer pH 5.0. SWV parameters: potential step 5 mV, frequency 260 Hz. The analyzed samples were deoxygenated prior to measurements by purging with argon (99.999%), saturated with water for 120 s. Isolation of nucleic acids proceeded on paramagnetic particles chemagic viral DNA, food DNA, DNA plant, DNA blood, and DNA tissue, all from Chemagen (USA). For automated isolation, apparatus epMotion 5075 from company Eppendorf (Germany) was used. Fully automated isolation was carried out on automated pipetting system epMotion 5075 (Eppendorf, Germany). The position of B4 is a magnetic separator (Promega). The positions of C1 and C4 can be thermostated (Epthermoadapter PCR96). The pipetting provides a robotic arm with adapters (TS50, TS300, TS1000) and Gripper (TG-T). The samples are placed in the position B3 in adapter Ep0.5/1.5/2ml. Module Reservoir is located in the position B1, where washing solutions and waste are available. The device is controlled by the epMotion control panel. The tips are located in the A4 (ePtips 50), A3 (ePtips 300) and A2 (ePtips 1000) positions. PCR 96 plates are used. The resulting volumes of collected samples ranged from 10 to 30 µl depending on the procedure.

Results and discussion

We created universal program by the help of software shipped with automatic pipetting station. We also optimized individual technical parameters and setting. Procedure was based on optimization, which was carried out at manual isolation. The most important parameters, which play crucial role in DNA isolation by MPs, are: i) pH of binding solution, in accordance with our results, the most suitable pH is 1,6; ii) ionic strength, which was in our experiments simulated by NaCl with final optimal concentration of 850 mM and iii) time of interaction between DNA and paramagnetic particles – for our experiments, the most suitable – optimal - interaction time is 15 min. In addition, we determined that with increasing intensity of shaking, increases also effectivity of DNA isolation in comparison with unshaken samples for more than 30%. Affectivity of DNA capturing by the help of epMotion was determined as 80%.

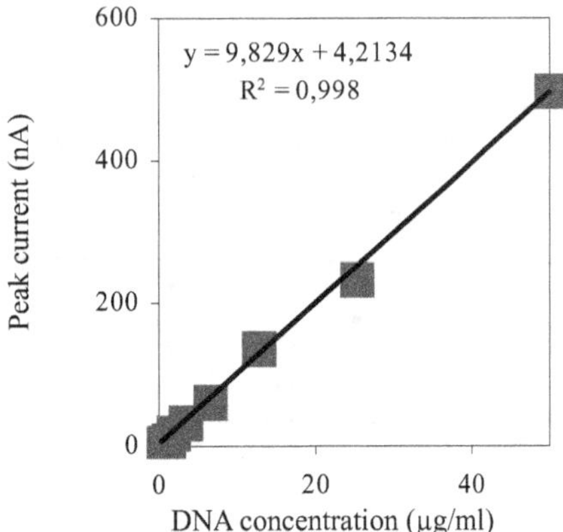

Figure 1: Dependence of DNA concentration on signal height. DNA was obtained by fully automated procedure.

Conclusion

Isolation of DNA can be carried out fully automatically, selectively and sensitively, with effectivity higher than 80 %, up to 60 min by the help of paramagnetic particles.

Acknowledgements

The work has been supported by grants: NANSEMED GA AV KAN20813081, NANIMEL GACR 102/08/1546, GACR 102/09/H083 and SIX CZ.1.05/2.1.00/03.0072.

References

Huska D, Hubalek J, Adam V, Kizek R (2008) Miniaturized electrochemical detector as a tool for detection of DNA amplified by PCR. Electrophoresis 29(24):4964-4971

Palecek E, Fojta M (2007) Magnetic beads as versatile tools for electrochemical DNA and protein biosensing. Talanta 74(3):276-290

Palecek E, Jelen F (2002) Electrochemistry of nucleic acids and development of DNA sensors. Crit Rev Anal Chem 32(3):261-270

Huska D, Hubalek J, Adam V, Vajtr D, Horna et al (2009) Automated nucleic acids isolation using paramagnetic particles. Talanta 79(2):402 - 411

Huska D, Adam V, Trnkova L, Kizek R (2009) The dependence of adenine isolation efficiency on the chain length evidenced using paramagnetic particles and voltammetry measurements. J Magn Magn Mater 321(10):1474-1477

J Biochem Tech (2010) 2(5):S89-S90
ISSN: 0974-2328

Single strand conformation polymorphism of oligonucleotides in their isolation by means of magnetic micro particles

Libuse Trnkova, Martina Fojtikova, Jana Bartoskova, Sylvie Holubova, Rene Kizek*

Received: 25 October 2010 / Received in revised form: 13 August 2011, Accepted: 25 August 2011, Published: 25 October 2011
© Sevas Educational Society 2011

Abstract

Studies of biomolecules depend strongly on efficiency of their isolation from biological materials. For these purposes the application of paramagnetic particles (MPs) represents a new method. This method, faster and less laborious than commonly used phenol chloroform extraction, is based on the hybridization of target molecules on a surface of magnetic particles. For the isolation of nucleic acids MPs are modified by certain sequences of thymine or other nucleotides. Using MPs we studied the isolation of ODNs (a) with different sequences of adenine and cytosine and (b)with different chain lengths. It was found that single–strand conformation polymorphism (SSCP) of ODNs is crucial phenomena for their isolation.

Keywords: Paramagnetic particles, oligonucleotides, electrochemical detection

Introduction

Study of biomolecules depends on the efficiency of their isolation from biological materials. There is permanent effort to find new procedures and technologies which will make this process more effective. The main requirements are rapidity and also quality and quantity of isolated, often very expensive, material. For isolation of nucleic acids (NA) there were developed various technologies and approaches based on elimination of interacting proteins and subsequent precipitation with alcohol. These methods have relatively high yields but they are very time consuming and need high cost laboratory instruments. Paramagnetic beads (MPs– magnetic particles) belong to the one of the new and promising methods for NA isolation. This method has a great advantage of designing of MPs size and surface, by which we can effectively isolate various molecules. Moreover, there is no complicated and laborious pre–treatment of a sample which significantly reduces the time needed to obtain the target molecules and reduced the risk of samples contamination. The method is non–invasive and is based on the physicochemical properties of paramagnetic particles. It consists of three steps: binding of nucleic acids on the surface of MPs, washing and releasing the target molecule and its detection. We focused on the study of isolation of oligonucleotides (ODNs) using magnetic particles according to the nucleotide sequence and the ODN chain length. ODN were detected electrochemically (Biyani and Nishigaki 2005; Cantor, Warshaw et al. 1970).

Materials and methods

Samples of oligonucleotides (Thermo Fisher Scientific, Ulm, Germany; Table 1) were prepared from lyophilized oligonucleotides (ODN), concentration was 50 μg.ml^{-1} (content of adenine was the same for all ODNs).

Their concentration was determined spectrophotometrically (Spectronic Unicam, England). The surface of paramagnetic particles (Invitrogen Dynal AS, Norway) was modified sequences dT25. The size of the particles was 2.80 ± 0.2 μm. MPs were washed 3 times with buffer I (0.1 M NaCl, 50 mM Na$_2$HPO$_4$, 50 mM NaH$_2$PO$_4$). The washing was followed with hybridization. The MPs was added to hybridization solution, phosphate buffer and sample ODN. This mixture was placed on the centrifuge (Multi–spin MSC–3000, Biosan, Lithuania) with occasionally shaking and centrifugation. The hybridization occurred on the base pairing between adenine and the ODN sequences dT25 on the surface of MPs. The MPs with immobilized ODN were concentrated by using a magnetic stand; the hybridization solution was removed and followed by washing. Then the MPs were added to phosphate buffer II, and samples were placed on thermoblock (Thermo mixer 5355 Comfort/Compact, Eppendorf, Germany). The releasing of ODN

Libuse Trnkova, Sylvie Holubova

Faculty of Science, Kotlářská 2, CZ–611 37 Brno, Czech Republic

Martina Fojtikova, Jana Bartoskova

Department of Biochemistry, Faculty of Science, Kotlářská 2, CZ–611 37 Brno, Czech Republic

Rene Kizek

Department of Chemistry and Biochemistry, Faculty of Agronomy, Mendel University in Brno, Zemědělská 1, CZ–613 00 Brno, Czech Republic

*Tel: +420 545 133 350, Fax: +420 545 212 044
E-mail: kizek@sci.muni.cz

from the magnetic particles occurred at 85°C. The last step was to re–concentration of MPs and pipetting the sample with an isolated ODN away. For the electrochemical detection a cell with three electrodes (the reference electrode – Ag/AgCl/3M KCl, auxiliary electrode – carbon , working electrode – a hanging mercury drop electrode with an area of 0.4 mm^2 -HMDE) was used.

Table 1: Characterization of ODNs (A – adenine, C- cytosine)

ODN	Number of A or ODN sequence	λ[cm^2 mmol^{-1}]	Molecular weight [g.mol^{-1}]	Melting point [°C]
dA9	9	12378	2757.9	18.0
dA12	12	12283	3697.5	24.0
dA18	18	12189	5576.8	37.6
dA36	36	12094	11214.6	60.1
Poly(dA)	400	12009	400000	73.2
H9 double	ACC CAC CCA ACC CAC CCA	9078	5288.5	57.7
H10 double	CCC CCC AAA CCC CCC AAA	8933	5288.5	60.9
H11 double	CCA CAC ACC CCA CAC ACC	8944	5288.5	51.8

Supporting electrolyte was acetate buffer with pH 4.5 (0.2 M CH_3COOH, 0.2 M CH_3COONa). The procedure of adsorptive transfer stripping was chosen. Voltammetric measurements were carried out with AUTOLAB analyzer (EcoChemie, the Netherlands) in connection with VA–Stand 663, (Metrohm, Switzerland). ODNs were also determined with increasing accumulation time (5, 10, 30, 60, 120, 240 and 360 s).

Results and discussion

Reduction signals of adenine (A) and cytosine (C) in ODNs were detected by linear sweep voltammetry associated with adsorptive transfer stripping (AdTS) technique (Huska Adam et al. 2009; Huska, Hubalek et al. 2009). The dependence of heights of reduction peaks on the accumulation time was investigated. ODNs were accumulated on the mercury electrode from 5 s to 360 s. We studied the signals of A, respectively A + C before and after isolation of ODNs by magnetic beads. The values of reduction peaks were measured with a polarization rate 100 mV.s^{-1} at pH 4.5. The interesting observation is that reduction signals of C and A depend strongly on chain length and sequence of nucleobases for ODNs before and after isolation by MPs. Therefore, the isolation is affected by single–strand conformation polymorphism SSCP (Biyani and Nishigaki, 2005; Cantor et al., 1970; Warshaw and Tinoco, 1966). The highest value of the reduction peak was detected at accumulation time of 120 s. According to reduction signals it can be suggested a similar structure of dA9 and dA12 compared to dA18 and dA36 (Fig 1). ODN H10 with double sequence (Table 1) provides significantly low reduction signal. This phenomenon can be justified by the sequence of six Cs and three As which may non–covalently interact and create the i–motif. Bases are therefore less accessible to the reduction onto the electrode surface. Sequence H9 double and H11 double are mirror arranged. Higher signal for H9 can be explained by this ODN sequence ends with adenine, which is protonized at pH 4.5 and better interact with the negatively charged electrode.

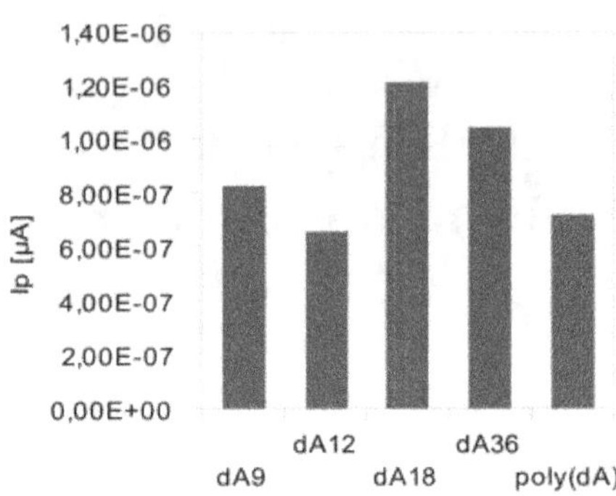

Figure 1: Reduction peak heights of adenine [A] for homo–ODNs isolated by MPs (pH 4.5, scan rate 800 mV.s–1, potential step 2mV)

Conclusion

The interesting study of single–strand conformation polymorphism (SSCP) and its effect on the isolation by MPs is novel. It was found that a chain length and a single nucleotide difference in ODNs can affect the change in their:

- Structure in buffer solutions
- Interaction with mps modified by dt25
- Dynamics on electrically charged interface
- Electrochemical signals
- Electrophoretical mobility

The effectiveness of isolation is influenced by MPs and ODN structure and possibly changing the structure during the isolation process. Electrochemical experiments revealed dynamic structures of ODN molecules in the dependence of their lengths and nucleotide sequences.

Acknowledgement

This work was supported by the following grants: INCHEMBIOL MSM0021622412, MSM0021630503 (MICROSYN), BIO–ANAL–MED LC06035 from the Ministry of Education, Youth and Sports of the Czech Republic and GAAV grant KAN208130801 (NANOSEMED).

References

Biyani M, Nishigaki K (2005) Structural characterization of ultra-stable higher-ordered aggregates generated by novel guanine-rich DNA sequences. Gene 364:130-138

Cantor CR, Warshaw MM, Shapiro H (1970) Oligonucleotide interactions .3. Circular dichroism studies of conformation of deoxyoligonucleotides. Biopolymers 9(9):1059-1077

Huska D, Adam V, Trnkova L, Kizek R (2009) Dependence of adenine isolation efficiency on the chain length evidenced using paramagnetic particles and voltammetry measurements. Journal of Magnetism and Magnetic Materials 321(10):1474-1477

Huska D, Hubalek J, Adam V, Vajtr D, Horna A et al. (2009) Automated nucleic acids isolation using paramagnetic microparticles coupled with electrochemical detection. Talanta 79(2):402-411

Warshaw MM, Tinoco I (1966) optical properties of 16 dinucleoside phosphates. Journal of Molecular Biology 20(1):29-38

J Biochem Tech (2010) 2(5):S91-S93
ISSN: 0974-2328

Box counting method in recording, processing and evaluation of genomic signals

Martin Valla*, Jiri Nedved, Dusan Pavlik, Vojtech Adam, Jaromir Hubalek, Libuse Trnkova, Rene Kizek, Ivo Provaznik

Received: 25 October 2010 / Received in revised form: 13 August 2011, Accepted: 25 August 2011, Published: 25 October 2011
© Sevas Educational Society 2011

Abstract

Fractals are geometrical shapes with noninteger dimension and invariance against change of scale factor. For each geometrical shape, a single parameter - dimension - can be calculated. For calculation, Box Counting Method (BCM) was chosen. Determination of dimension on one level of resolution is not sufficient, it is necessary to subsequently process it and determine multifractal coefficient. Aim of our interest consists in analysis of images obtained from a linear sequence of DNA code. For analysis itself, sequences of *Homo sapiens* and *Citrobacter youngae* were chosen. Results of calculation are fractal coefficients derived from dimensions of generated structures. This result enables to introduce criterion for sequences determination. Graphical outputs may be also represented as multidimensional alternative transformation of linearly recorded genomic signal. Algorithms were developed in computational environment MATLAB. Data were downloaded from a public database EMBL (ESI) and GenBank (NCBI).

Keywords: Fractals, dimension, *Citrobacter youngae, Homo sapiens*

Martin Valla*, Jiri Nedved, Dusan Pavlik, Ivo Provaznik

Department of Biomedical Engineering, Faculty of Electrical Engineering and Communication, Brno University of Technology, Kolejni 4, CZ-61200 Brno, Czech Republic

*Tel: +420 541 149 411, Fax: +420 541 211 697
  E-mail: valla@phd.feec.vutbr.cz

Vojtech Adam, Rene Kizek

Department of Chemistry and Biochemistry, Faculty of Agronomy, Mendel University in Brno, Zemedelska 1, CZ-613 00 Brno, Czech Republic

Jaromir Hubalek

Department of Microelectronics, Faculty of Electrical Engineering and Communication, Brno University of Technology, Udolni 53, CZ-602 00 Brno, Czech Republic

Libuse Trnkova

Department of Chemistry, Faculty of Science, Masaryk University, Kotlarska 2, CZ-611 37 Brno

Introduction

Theory of fractals originated on the basis of observation of natural structures. Application of this knowledge for DNA analysis leads to additional insight to genome. Term *fractal* is derived from latin word fractus, which means fraction. Fractals are geometrical shapes, which have non-integer dimension and are similar for each other. Self-similarity is a phenomenon, which occurs in case, when structure looks identical at whatever magnification. Mathematically, this phenomenon is called invariance against change of scale factor. For each object, its dimension can be calculated. There are many approaches for calculation of this magnitude. For calculation, Box Counting Method (BCM) method was chosen. BCM demonstrates simplicity and lucidity, but on the other hand, disadvantage in processing of colour, or black and white textures are evident. Determination of dimension on one level of resolution is insufficient; calculation must be necessarily carried out in its entirety. Resulting dimensions must be subsequently processed in accordance with chosen procedure; multifractal coefficient must be subsequently determined.

Aim of our object consists in application of theory of fractals for analysis of images obtained from sequence of DNA code. Sequences encoding two proteins (*Homo sapiens* and *Citrobacter youngae*) were chosen for our analysis. Results are fractals coefficients derived from dimensions of structures. They enable introducing of criterion for sequences differentiation. Output can be also presented as multidimensional transformation – imaging – and record of genomic signal (from linear entry in line to planar in Fig 1). All algorithms were developed in computational environment MATLAB.

Materials and methods

Image presentation of DNA

Linear record for DNA sequence (for example AGGCTGGAATGC) must be transformed into Figure 1 by following procedure. Square with four apexes - A, C, G, T (opposite pairs A-C and G-T) - and initial point, which is situated in the centre of square, are defined. The square is determinically divided by points by modified method of *Chaos Game*. The method defines an initial point connected with apex, which is given by character at given position in sequence and at one half of distinct between initial point and apex, point is projected; new point is initial for new interaction – a new connection of point with next square apex is given by subsequent character of sequence; schematic demonstration in Berthelsen 1993. First four steps of both methods are demonstrated in Figure 1.

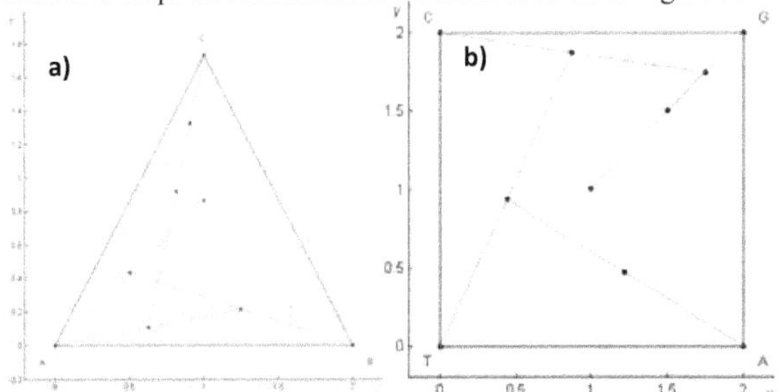

Figure 1: (a) A model example of generation of attractor of Sierpinski triangle by Chaos Game method (b) initiation of attractor designed from first four nucleotides of sequence GCTA

Fig 2a represents information carrying English definition "*Homo sapiens*, Cu++ transporting, alpha polypeptide (ATP7A) on chromosome X'" marked with locus NG_013224 and with length of 146 699 bp. Fig 2b is created from sequence "*Citrobacter youngae* ATCC 29220 C_sp-1.0.1_Cont0.7, whole genome shotgun sequence" marked with locus NZ_ABWL02000007 and with length of 274 831 bp. Lengths of sequences (bp) determine numbers of points marked in square.

ATP7A

Description of sequence „Homo sapiens ATPase, Cu++ transporting, alpha polypeptide (ATP7A) on chromosome X" is given by locus NG_013224 and length 139 699 bp. This sequence encodes transmembrane protein, which enables transport of copper(II) ions through cell membrane. A base is represented by 28,9 %, C base by 19,3 %, G base by 19,4 % and T base by 32,5 %.

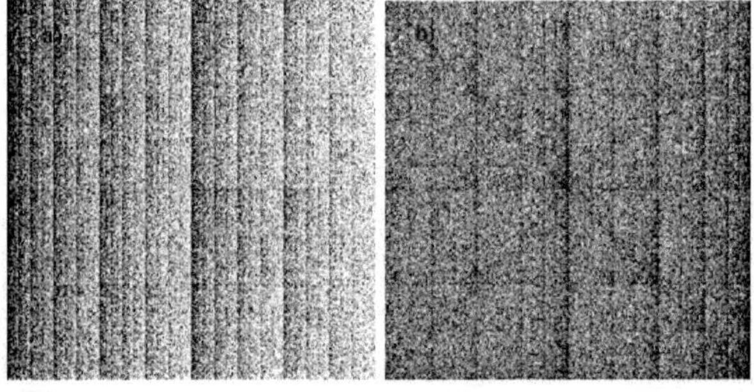

Figure 2: point presentation of part of DNA codes. On the left (a) is sequence NG_013224 and on the right (b) is sequence Z_ABWL02000007
ATCC 29220

Sequence is in database described as "*Citrobacter youngae* ATCC 29220 C_sp-1.0.1_Cont0.7, whole genome shotgun sequence" and is characterized by locus NZ_ABWL02000007 and length 274 831 bp. Sequence originates in genome of bacterium *Citrobacter youngae* strain ATCC 29220. Sequence contains 23,7 % of A base, 27,7 % of C base, 25,2 % of G base and 23,4 % of T base.

For more well-arranged illustration, NG_013224 can be projected in colour by association with one colour to one nucleotide - A (yellow) C (green) G (blue) T (red). See Figure 3a in black and white presentation, Figure 3b of the same sequence in colour presentation.

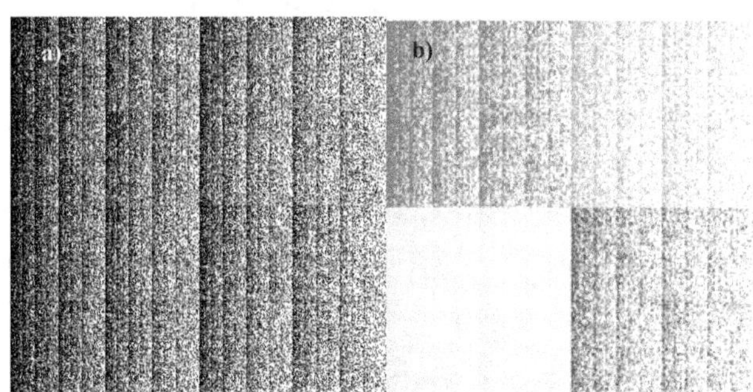

Figure 3: Point presentation of DNA codes of sequence NG_013224. On the left (a) in black and white presentation, on the right (b) in colour presentation of points depiction - G-blue, C-green, T-red, A-yellow.

After transformation of one-dimensional sequence into image, it is necessary to process this image in next step. Image presentation itself represents parameter, so image presentations can be subjectively compared with each other (by insight) on various level of approximation (Berthelsen, 1993). Next possibility consists in mathematical description and evaluation of image.

Box Counting Method processing

Box Counting Method was used for image analysis. Method is based on multiple calculation of classical Hausdorff dimension. Basic formula for calculation is in the form (Turner 1998)

$$D = \frac{\log N}{\log \frac{1}{r}} \qquad (1)$$

Where N is number of self-similar areas, r is size of area and D has significance as dimension of object.

BCM is based on principle of points summation below given mask. Size of mask determines scale of imaging. For each approximation, size of dimension is calculated according formula (1). For higher number of approximations, more dimensions can be obtained; these dimensions can be interlaid by line at imaging in graph. Direction of line is parameter of image and is called multifractal coefficient. Equation (1) can be modified as

$$\log N = D \cdot \log \frac{1}{r} \qquad (2)$$

This form also represents universal equation of line

$$y = k \cdot x \qquad (3)$$

Where $y = \log N$ determines number of points below mask, $k = D$ sequence of dimensions (multifractal coefficient), $x = \log (1/r)$, reverse value of mask size (in pixels).

Results and discussion

Results of functions from Fig 2a and Fig 2b are demonstrated in Figure 4. Dimensions are projected dextrosinistrally so that point on high right corresponds to first approximation (size of mask 2 x 2 pixels), point on the left to this point corresponds to second approximation (size of mask 3 x 3 pixels), etc. Direction of line, which interlays points after 24 approximations, is the searched multifractal coefficient in Figure 4.

After summarizing of fractal coefficients in table, result is following:

Table 1: Multifractal coefficients of two model sequences calculated by BCM method

ID sequence	D_{BCM} [-]
NG_013224	1.1445
NZ_ABWL02000007	0.9562

Conclusion

BCM provides calculating of dimensions on different levels of approximation and obtains resulting multifractal coefficient, which serves as parameter for mathematical description of given sequence (direction of line interlaying obtained dimensions). Fractal coefficient can be used for parameterization and mutual comparing of DNA sequences.

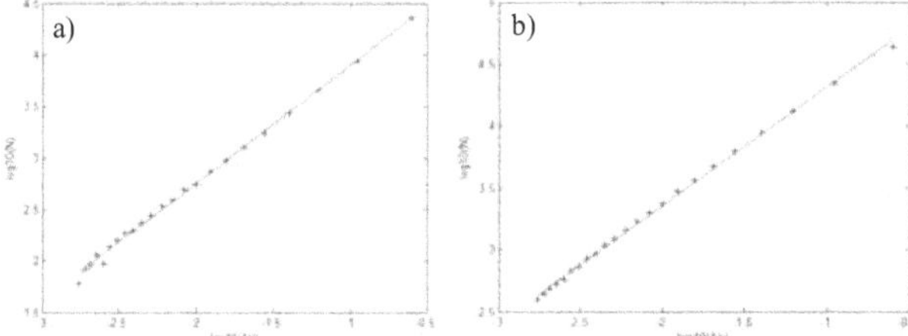

Figure 4: Result of BCM. (a) Result of BCM with multifractal coefficient 1.1445 [-] for sequence NG_013224, and (b) result of BCM with multifractal coefficient 0.9562 [-] for NZ_ABWL02000007.

Acknowledgement

This work was supported by grants FRVS TO A/2894, GACR 102/09/H083, GA AV KAN208130801, and LC06035.

References

Berthelsen ChL (1993) Fractal analysis of DNA sequence data. The University of Utah

Turner, JT (1998) Fractal Geometry in Digital Imaging, Leicester, Academic Press, ISBN 0-12-703970-8

J Biochem Tech (2010) 2(5):S94-S95
ISSN: 0974-2328

Utilization of FIA-UV/ED for detection of adenine derivates

Ondrej Zitka, Libuse Trnkova, Frantisek Jelen, Vojtech Adam, Rene Kizek*

Received: 25 October 2010 / Received in revised form: 13 August 2011, Accepted: 25 August 2011, Published: 25 October 2011
© Sevas Educational Society 2011

Abstract

A purine derivative adenine poses many biological functions. Besides the fact that this molecule is one of the building blocks for RNA and DNA, there are many derivates with their specifics attributes. 2-aminopurine is well known as mutagen. 2,6-diaminopurine is able to replace purine basis in nucleic acids. Benzylaminopurine belongs to phytohormones.

Keywords: Purine, adenine, oxidation, DNA, electrochemical determination

Introduction

Biological harmfulness of reactive oxygen species is given by the subsequent oxidation of essential cellular structures. They can therefore peroxide lipids to form hydrocarbon radicals and thus alter the structure and function of biomembranes. In the case of proteins, the amino acid oxidation, cleavage peptide linkages and other changes in the structure, function and protein-protein interactions occur. Nucleic bases (purines and pyrimidines) can be also oxidized by reactive oxygen species. The oxidized bases are than removed from the DNA chain. Unfortunately, all these changes may give rise to single-strand and double-strand breaks in DNA leading to

Ondrej Zitka, Vojtech Adam, Rene Kizek*

Department of Chemistry and Biochemistry, Faculty of Agronomy, Mendel University in Brno, Zemedelska 1, CZ-613 00 Brno, Czech Republic.

*Tel: +420 545 133 350, Fax: +420 545 212 044
E-mail: kizek@sci.muni.cz

Libuse Trnkova

Department of Chemistry, Faculty of Science, Masaryk University, Kotlarska 2, CZ-611 37 Brno, Czech Republic

Frantisek Jelen

Laboratory of Biophysical Chemistry and Molecular Oncology, Institute of Biophysics, v.v.i, Academy of Sciences of the Czech Republic, Kralovopolska 135, CZ-612 65 Brno, Czech Republic

irreversible cell damage. It has been demonstrated that molecule, which shows very well on DNA oxidation, is the creation of 8-oxo-2-deoxyguanosine (8-oxo-DG). Besides oxidation of guanine the oxidation changes in the molecules adenine (8-oxo-2-deoxyadenosine) were also observed (Quinlivan and Gregory 2008; Singh et al. 2009; Zhou et al. 2008). For the detection of this marker of oxidative damage number of methods (liquid chromatography, capillary electrophoresis) was developed. Electrochemical determination of these biological markers appears to be advantageous for its sensitivity and selectivity.

Materials and methods

In this study we aimed at study of electrochemical behaviour of derivates of adenine (adenosine-monophosphate, cyclic adenosine monophosphate, adenosine-triphosphate, 2-aminopurine, adenine, nicotinamide adenine dinucleotide, 2,6-diaminopurine, adenosine, 6-benzyl-aminopurine, S-adenosyl-L-Methionine). For this purpose the technique of flow injection analysis with electrochemical detection (FIA-ED) was employed. The FIA-ED system was consisting of one solvent-delivery pump, injection valve. Two serially connected detectors in tandem were presented. First was UV detector which was set on 260nm wavelength and second one was Analytical electrochemical cell (5040, ESA, USA) which is consisted of glassy carbon working electrode, hydrogen-palladium electrode as reference electrode and auxiliary electrode, and Coulochem III as a control module. All adenine derivates were diluted in the Mili Q water. 10 µl of sample was injected in the system. Samples of matrixes were centrifuged at 14 000 G by time of 20 minutes. Supernatant was then directly analyzed.

Results and discussion

First of all, we optimized detection method, where hydrodynamic voltammograms within the range from 100 to 1,300 mV for all above mentioned analytes were measured. Based on the obtained results, we chose 1.000 mV as suitable for sensitive detection of all derivates of nucleic acids bases. Influences of pH and flow rate on signal height were also tested. pH optimum was 5 and suitable flow rate was determined as 0.75 ml.min^{-1} (Fig 2). Dose-response curves were measured within the range from 1 to 100 µM for adenine and 2-aminopurine; for other analytes, concentration range from 1 to

1,000 µM was chosen. Moreover, we investigated behaviour of adenine and its derivatives in the presence of two types of matrices.

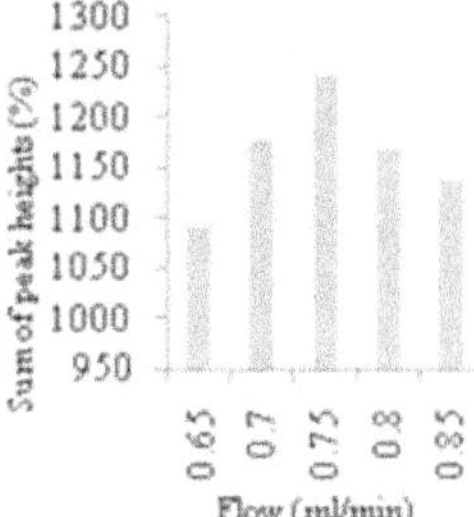

Figure 1: Dependence of peak height of adenines on flow rate.

Human urine and extract of BY-2 tobacco cells were chosen as suitable biological matrices. Obtained recoveries for each analytes demonstrate various interactions with these two matrices. Samples of matrixes were first injected to obtain one electrochemical signal under 1000mV which indirectly indicates the concentration of adenine derivates. After spiking of matrix by each adenine derivate we obtained information about approximate ability of matrix to scavenge of particular derivative. While the peak of spike+sample is higher in comparison to only sample peak the ability of matrix to scavenge particular derivate is theoretically low (Fig 2).

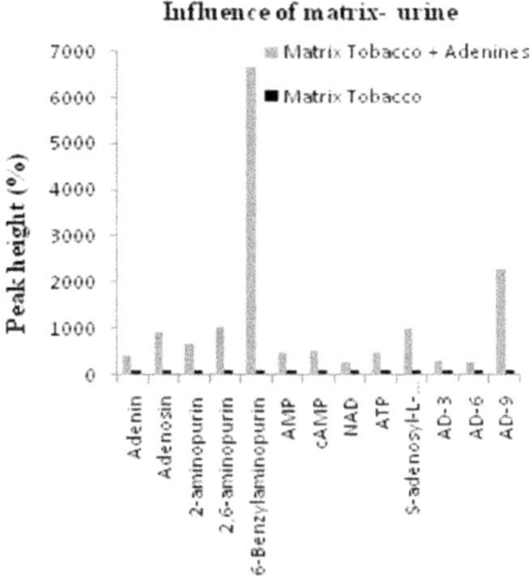

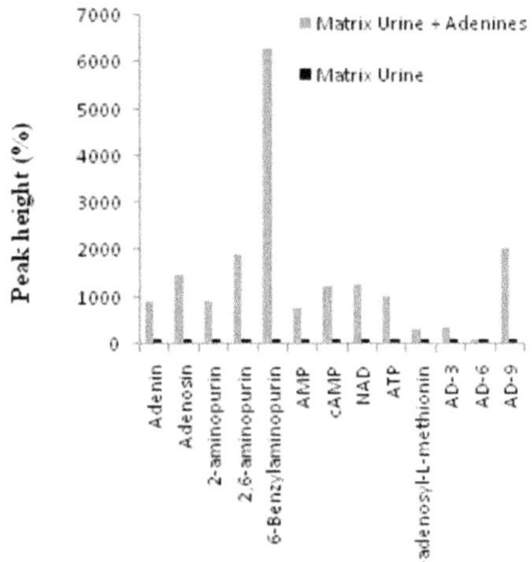

Figure 2: Influence of matrix urine & tobbaco

Conclusion

Liquid chromatography represents so called "golden standard" for analysis of complex mixtures and matrices. Its connection with electrochemical detector brings higher sensitivity and selectivity.

Acknowledgements

The work has been supported by GA AV KAN208130801, GA ČR 102/08/1546 a GAČR 102/09/H083.

References

Quinlivan EP, Gregory JF (2008) DNA methylation determination by liquid chromatography-tandem mass spectrometry using novel biosynthetic [U-N-15]deoxycytidine and [U-N-15]methyldeoxycytidine internal standards. Nucleic Acids Research 36(18):119

Singh R, Teichert F, Verschoyle RD, Kaur B, Vives M, et al (2009) Simultaneous determination of 8-oxo-2 '-deoxyguanosine and 8-oxo-2 '-deoxyadenosine in DNA using online column-switching liquid chromatography/tandem mass spectrometry. Rapid Communications in Mass Spectrometry 23(1):151-160

Zhou JL, Qian ZM, Luo YD, Tang D, Chen H, et al (2008) Screening and mechanism study of components targeting DNA from the Chinese herb Lonicera japonica by liquid chromatography/mass spectrometry and fluorescence spectroscopy. Biomedical Chromatography 22(10):1164-1172

J Biochem Tech (2010) 2(5):S96-S97
ISSN: 0974-2328

Multi-instrumental studying of interaction between heavy metal ions and free aminoacids

Ondrej Zitka, Jiri Sochor, Dalibor Huska, Zdenek Janicek, Dusan Pavlik, Martin Valla, Radim Hrdy, Natalia Cernei, Jaromír Hubalek, Vojtech Adam, Ales Horna, Ivo Provaznik, Rene Kizek*

Received: 25 October 2010 / Received in revised form: 13 August 2011, Accepted: 25 August 2011, Published: 25 October 2011
© Sevas Educational Society 2011

Abstract

The main contribution of this work is in use of different techniques for the study of interactions in the field of metallomics. Interactions of free amino acids with heavy metals are subjects, which have only partial scientific interest. However, both the transport of metals and metal binding with free amino acids and/or small molecules have been still only poorly understood. We developed few methods spectrophotometric, flow injection analysis and chromatographic both with various type of electrochemical detection, which all were successfully used for studying of creation of complexes between cadmium or platinum derivatives *in vitro*.

Ondrej Zitka, Dalibor Huska , Natalia Cernei, Vojtech Adam, Rene Kizek

Department of Chemistry and Biochemistry, Faculty of Agronomy, Mendel University in Brno, Zemedelska 1, CZ-613 00 Brno, Czech Republic.

*Tel: +420 545 133 350, Fax: +420 545 212 044
E-mail: kizek@sci.muni.cz

Jiri Sochor

Department of Fruit Growing, Faculty of Horticulture, Mendel university in Brno, Valticka 337, 691 44 Lednice.

Zdenek Janicek, Dusan Pavlik, Martin Valla

Department of Biomedical Engineering, Faculty of Electrical Engineering and Communication, Brno University of Technology, Kolejni 4, CZ-612 00 Brno, Czech Republic.

Radim Hrdy, Jaromír Hubalek, Ivo Provaznik

Department of Microelectronics, Faculty of Electrical Engineering and Communication, Brno University of Technology, Technick a 2896/2, CZ-616 69 Brno, Czech Republic.

Ales Horna

Radanal Ltd, Okruzni 613, CZ-530 03, Pardubice, Czech Republic,

Keywords: aminoacids, heavy metals, interaction, metallothionein fragment, HPLC with electrochemical detection.

Introduction

Biologically active metals can be included in two groups - essential and toxic with respect to used physiological criteria. Free amino acids (AA) are AA, which are not bounded in proteins or in other biochemically active structures. Furthermore, we know also biogenic amines, which are originated from biogenic AA or they are their precursors. In the case that toxic metals can occur in free ionic state in the organism, they are able to generate reactive oxygen species (ROS). ROS can subsequently oxidatively damage number of biomolecules and thus disrupt the homeostasis. In first line of defence against ROS – their detoxification, small biological active molecules, which are able to scavenge ROS, play crucial role. Cysteine with thiol group the most importantly participates on ROS scavenging, together with nucleophile centres of side chains of amino acids methionine, histidine of tryptophan.

Materials and methods

Aim of this work consisted in *in vitro* monitoring of interactions between metals and AA by use of different techniques. Experiments were carried out on several levels. Interaction between free sulphur containing AA and cadmium, between cysteine-rich peptides and cadmium and platinum based cytostatics and between protein metallothionein (MT) and metal were investigated.

Results and discussion

Observing of interaction between AA and cadmium was carried out thanks to monitoring of absorption maximum of Murexide (Leverrier et al. 2007). Furthermore there was observed creation of complex due to factorial analysis with change of concentrations of both reagents and time of interaction. Samples prepared for studying of intensity of interaction were carried out using automated preparation of sample with robotic station and so detection by Ellman reaction in automatic analyser. This high throughput approach was designed because many of parameters (concentration of components, various temperature or various time of interaction) had to be studied. For determination of intensity of interaction between peptides and metals a flow injection analysis with electrochemical detection (FIA-ED) was employed. FIA-ED method was optimized for detection of cysteine-rich peptides from MT and Platinum cytostatics by amperometric detection. For this method a glassy carbon (GC) electrode as working was employed and applied potential 500mV was used with regards to interference of cisplatin. After we obtained the best parameters (flow rate applied potential, electrolyte) we determined the intensity of interaction by determination of decreasing of free thiol groups which peptides contained. After high complex creation we detected fast decreasing slopes and that shows which of tested peptides was the best for interaction. Than the same method was coupled with coulometric detection instead of amperometric and used for detection of peptides in presence of cadmium. The most effective applied potential was 900mV. Obtained results after correlation gave information about creation of complex (Zitka et al. 2007). Same as in previous case the slopes of decreasing of signals of thiol groups defined of complex creation by all studied peptides.

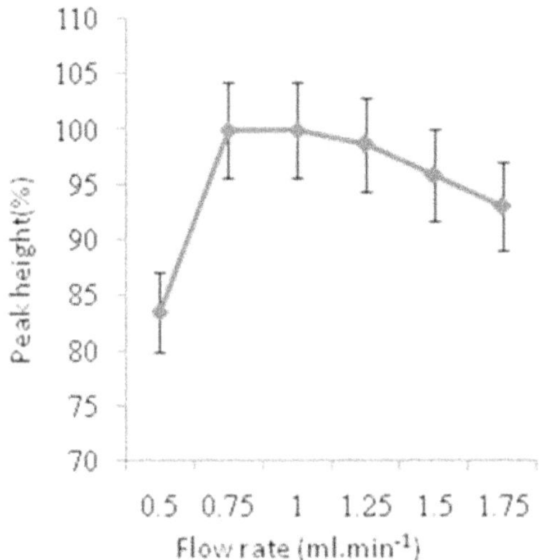

Figure 1: influence of flow rate on peak height of separately analyzed fragments of MT.

After connection of chromatographic column Kinetex C-18 (150 × 4.6; 2,5 μm beads, Phenomenex, USA) into the gradient system (mobile phase: A: 80mM TFA, B: Methanol) with coulometric detection we developed effective separation of peptides for further studies of interactions between them and strong ligands. Gradient was linearly increasing from 3%B to 50%B in 20 minutes and 20μl of the sample was injected to the system where flow rate was 1ml/min. Optimal detection was under applied potential 900mV. We used this method might be very useful for basic competitive studying of complex creation metal could be added to the mixture and after incubation and subsequent separation the chromatogram compared to the untreated control variant can provide information as decreasing which is associated with complex creation or complex could change the retention time also. Finally for observing of complex of MT with metals we used the QCM system in optimised conditions pH 8 and time of interaction 180s. MT was immobilised on gold QCM crystal and than analyzed by electron microscopy. There were visible bright regions which represented metal occurrence in MT domains (Adam et al. 2007).

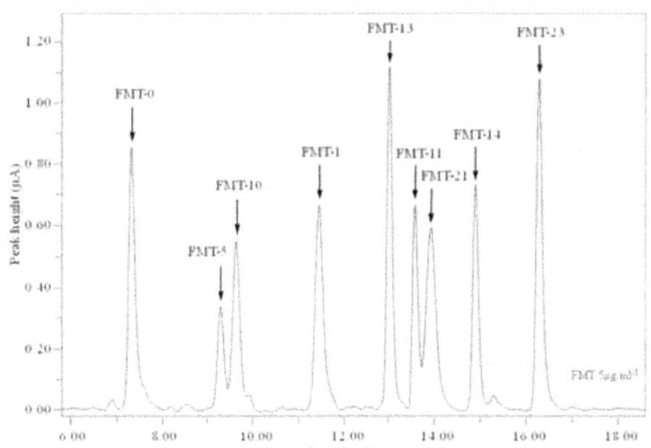

Figure 2: Chromatogram of various peptide fragments of MT acquired by HPLC-ED.

Conclusion

Methods developed and used in this work can be very useful in solving of presented questions and represent alternative to more costly analytical techniques, such as NMR or X-ray analysis. In addition to application possibilities of UV-Vis spectroscopy and advanced flow electrochemical detectors, this work covers field of proteomics. Our approach demonstrates complementarily of proteomics and metallomics. The protein sequence of the specific protein can be used *in vitro* for advanced studies, which are the aim of interest of metallomics. *In vitro* studies of metallothionein protein fragments can provide also reference data that could confirm or disconfirm computational analysis of complex formation.

Acknowledgements

The work has been supported by GA AV KAN208130801, GA ČR 102/08/1546 a GAČR 102/09/H083.

References

Adam V, Krizkova S, Zitka O, Trnkova L, Petrlova J, et al (2007) Determination of apo-metallothionein using adsorptive transfer stripping technique in connection with differential pulse voltammetry. Electroanalysis 19(2-3):339-347

Leverrier P, Montigny C, Garrigos M, Champeil P (2007) Metal binding to ligands: Cadmium complexes with glutathione revisited. Analytical Biochemistry 371(2):215-228

Zitka O, Huska D, Krizkova S, Adam V, Chavis GJ, et al (2007) An investigation of glutathione-platinum(II) interactions by means of the flow injection analysis using glassy carbon electrode. Sensors 7(7):1256-1270

J Biochem Tech (2010) 2(5):S98-S99
ISSN: 0974-2328

Catalytic hydrogen evolution by polyaminoacids using mercury electrode

Marko Živanovič, Veronika Ostatná, Emil Paleček*

Received: 25 October 2010 / Received in revised form: 13 August 2011, Accepted: 25 August 2011, Published: 25 October 2011
© Sevas Educational Society 2011

Abstract

It was shown that using constant current chronopotentiometric stripping (CPS) peptides and proteins at nanomolar concentrations produce protein structure–sensitive peak H at mercury electrodes. This peak is due to the catalytic hydrogen evolution reaction (HER). Polyamino acids can be considered as an intermediate model system between peptides and macromolecular proteins. Here we used polyamino acids (poly(aa)) such as polylysine (polyLys) and polyarginine (polyArg) and cyclic voltammetry or CPS in combination with hanging mercury drop electrode to explore how different amino acid residues in proteins contribute to the catalytic HER.

Keywords: constant current chronopotentiometry, mercury electrodes, polyaminoacids, electrocatalysis

Introduction

First paper on electrochemistry of proteins was published by Heyrovský and Babička in 1930 describing the d.c. polarographic "presodium wave". Soon it was shown that cysteine–containing proteins produce well–developed Brdička's double wave (Brdicka's catalytic response, BCR) (Brdicka, 1933). Both the presodium wave and BCR were shown to be due to the catalytic hydrogen evolution. Recently, we have shown that using CPS with hanging mercury drop electrode (HMDE), peptides and proteins yield at negative potentials a well developed peak (peak H) in buffered solutions, regardless of the presence or absence of cobalt (Palecek and Ostatna, 2007; Tomschik et al., 1998). Here we show that and polyArg produced CPS peak H at low bulk concentrations at highly negative potentials due to HER.

Marko Zivanovic, Veronika Ostatna, Emil Palecek[*]

Institute of Biophysics, Academy of Sciences of the Czech Republic, Královopolská 135, 612 65 Brno, Czech Republic

*Tel: +420 541 517 177, Fax: +420 541 21 12 93
E-mail: palecek@ibp.cz

Materials and methods

Polylysine, MW = 4000 – 15000 (poly-L-Lys 4-15), polylysine MW = 30000 – 70000 (poly-L-Lys 30-70), and polyarginine MW = 5000 – 15000 (poly–L–Arg 5-15) were purchased from Sigma–Aldrich. In all adsorptive stripping experiments, the analyte was adsorbed on the hanging mercury drop electrode (HMDE) surface from the stirred solution containing poly(aa) solution in the background electrolyte at a given accumulation potential (E_A), during the chosen accumulation time (t_A); then chronopotentiograms or voltammograms were recorded.

Results and Discussion

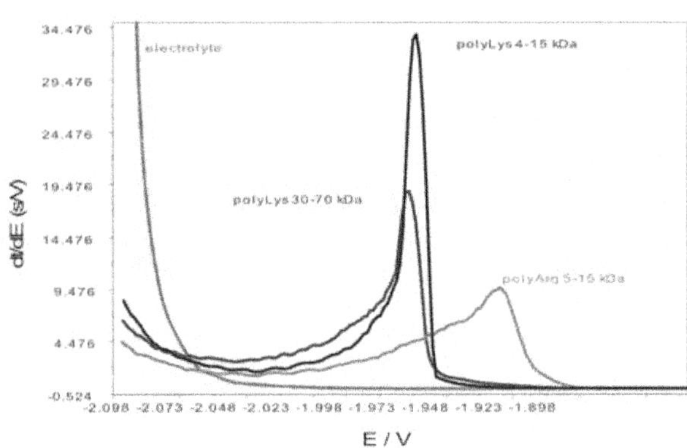

Figure 1: Adsorptive stripping CPS peak H of 0.5 µM poly–L–Lys 30–70, 0.5 µM poly–L–Lys 4–15, and 0.5 µM poly–L–Arg 5–15. Background electrolyte: 0.05 M McIlvaine buffer, pH 6. I_{str} = - 30 µA, t_A = 60 s, E_A = E_i = - 0.1 V, E_f = - 2.1 V, 23.7 °C. Poly(aa) concentrations are related to their monomer content.

Our results show that lysine or arginine residues in a polypeptide chain are sufficient to induce catalytic HER at the HMDE. We found that at pH 6 both polylysine and polyarginine produced comparable peaks H under the given conditions.

Conclusion

In recent years catalytic HER of proteins measured by CPS at HMDE and amalgam electrodes appeared as a new tool in protein analysis (Doneux et al. Palecek and Ostatna 2007). High sensitivity and low volume requirements (femto– and attomoles of proteins can be analyzed) as well as simple and inexpensive instrumentation made this method attractive for applications in proteomics and biomedicine. This work represents an attempt to apply poly(aa)'s to contribute to elucidation of catalysis of HER by proteins.

Acknowledgement

This work was supported by grants from the Academy of Sciences of the Czech Republic: KAN400310651 to EP and KJB100040901 to VO, MEYS CR ME09038 to EP, Research Centre LC06035 and the Institution research plans Nos. AV0Z50040507 and AV0Z50040702.

References

Brdička, R (1933) Collect Czech Chem Commun 5:148

Doneux T, Dorcak V, Palecek E (2007) Influence of the Interfacial Peptide Organization on the Catalysis of Hydrogen Evolution. Langmuir 26(2):1347-1353

Heyrovský J, Babička J (1930) Collect Czech Chem Commun 2:370

Palecek E, Ostatna V (2007) Electroactivity of nonconjugated proteins and peptides. Towards electroanalysis of all proteins. Electroanalysis 19(23):2383-2403

Tomschik M, Havran L, Fojta M, Palecek E (1998) Constant current chronopotentiometric stripping analysis of bioactive peptides at mercury and carbon electrodes. Electroanalysis 10(6):403-409

J Biochem Tech (2010) 2(5):S100-S101
ISSN: 0974-2328

Analysis of DNA modified by cerium (III), lanthanum (III) and gadolinium (III) ions by using of raman spectroscopy

Veronika Kohoutkova, Petr Babula, Radka Opatrilova, Oldrich Vrana, Vojtech Adam, Rene Kizek*

Received: 25 October 2010 / Received in revised form: 13 August 2011, Accepted: 25 August 2011, Published: 25 October 2011
© Sevas Educational Society 2011

Abstract

Lanthanides are at the present group of heavy metals, which are at the centre of interest, especially because of their ability to interact with DNA and similarity with calcium(III) ions, which play crucial role in many cell processes. On basis of DNA interactions, anti–cancerogenic potential is established. In our work, we were focused on interactions of chosen lanthanides – lanthanum, cerium and gadolinium – with DNA. Results of our experiments demonstrate ability of lanthanides to form DNA adducts.

Keywords: DNA, raman spectroscopy, coumarin complexes

Introduction

The biological properties of the lanthanides have stimulated research into their therapeutic application. Up–to–date, we have

Veronika Kohoutkova, Petr Babula

Department of Natural Drugs, Faculty of Pharmacy, University of Veterinary and Pharmaceutical Sciences, Palackeho 1–3, CZ–612 42 Brno, Czech Republic

Radka Opatrilova

Department of Chemical Drugs, Faculty of Pharmacy, University of Veterinary and Pharmaceutical Sciences, Palackeho 1–3, CZ–612 42 Brno, Czech Republic

Oldrich Vrana

Institute of Biophysics, Academy of Sciences of the Czech Republic, Kralovopolska 135, CZ–612 65 Brno, Czech Republic

Vojtech Adam, Rene Kizek*

Department of Chemistry and Biochemistry,Faculty of Agronomy, Mendel University of Agricultu re and Forestry, Zemedelska 1, CZ–611 37 Brno, Czech Republic

*Tel: +420 545 133 350, Fax: +420 545 212 044
E-mail: kizek@sci.muni.cz

been successfully using lanthanides(III) compounds for their anti–cancer potential. Cerium(III), lanthanum(III) and neodymium(III) coumarin complexes were synthesized with ligands, such as hymecromone, umbellipherone, mendiaxon, warfarin, coumachlor and niffcoumar. Preclinical studies with these compounds have demonstrated cytotoxicity against the HL–60 myeloid cell line. Clinical reports suggested that above mentioned compounds demonstrated cytotoxic effect; however the biochemical mechanism of their action is still unclear. Therefore, we aim our attention on studying of interactions DNA with inorganic salts of cerium(III), lanthanum(III) and gadolinium(III) (Duguid et al. 1995; Fricker 2006; Schmitt & Popp 2007; Vrana et al. 2007).

Materials and methods

Spectra of samples – DNA after interaction with individual lanthanide – were measured on the Raman spectrometer Jobin–Yvon T64000 (Horiba Scientific, Japan). Concentrated solutions of samples were sealed in the glass capillaries Kimax 34507 (Kimble products, USA) and placed into focused laser beam. Excitation wavelength radiation was 488 nm (argon laser Innova 90C Fred, Coherent). The intensity of radiation reaching the sample was 100 mW and was monitored by using of Broadband Power Meter (Melles Griot, USA). Data processing was done by software LabSpec. From all processed spectra, were deducted contributions of solvent (0.1 M NaCl). In the case of differential spectra, they were deducted from the modified spectrum of DNA. The final appearance of the spectrum was treated in the Origin 8.

Results and Discussion

A. Lanthanum. Raman spectrum shows that the ratio of 683/670 cm^{-1} belts is less than 1, which marks the transition from B–conformation in the A–DNA. The increasing intensity of the peak 750 cm^{-1} is connected with the vibration of lanthanum. The rise of the belt in the area of 804 cm^{-1} indicates the local structural changes in DNA modified by lanthanum. A significant decrease in the intensity of the band 1490 cm^{-1} corresponds to change in the position N7 of guanine in all cases of the measured samples. The rise of intensity for the belt 1578 cm^{-1} corresponds to the changes in the G–C and T–A pairs of the bases.

B. Cerium. Similarly to DNA modification by lanthanum(III) ions, there were noticeable changes in the areas of 670 cm^{-1} and 683 cm^{-1}. The rate of intensities of belts 683/670 cm^{-1} is 0.7, which means the most transition from B–DNA into A–DNA of the modified DNA by the studied ions. The belts in the area of 735 cm^{-1}, 920 cm^{-1}, 973 cm^{-1}, 1018 cm^{-1}, 1073 cm^{-1}, 1137 cm^{-1}, 1211 cm^{-1}, 1337 cm^{-1}, and 1448 cm^{-1} are connected with contribution cerium vibrations.

C. Gadolinium. The ratio of 683/670 cm–1 is less than 1. DNA modified by gadolinium(III) ions changed its modification from B–DNA into A DNA. The increase in the field of 807 cm^{-1} is the highest from all DNAs modified by the measured lanthanides(III) ions. The belts in the area of 1078 cm^{-1}, 1218 cm^{-1}, 1361 cm^{-1}, 1448 cm^{-1} are linked to the contribution of the vibration of gadolinium.

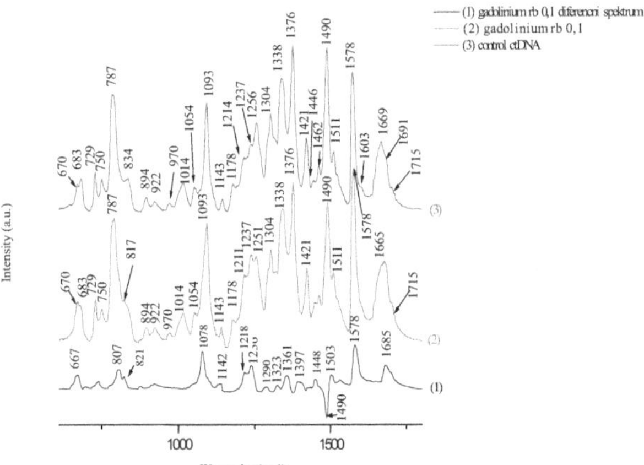

Figure 1: Changes of DNA spectra after interaction with gadolinium(III) ions. Typical belts are indicated by arrows. Changes in spectra are connected with formation of DNA adducts as well as with vibrations of gadolinium(III) ions themselves.

Conclusion

The differential Raman spectrum of DNA modified by all complexes is falling linearly in dependence on rb. For all measured modified DNAs, there are obvious changes in both parts of the spectra: vibration of the sugarphosphate skeletons as well as in the areas of vibration bases not least in the areas of the vibration of the individual ions (La, Gd, and Ce). For DNA modified by all ions there is a noticeable shift from B–DNA into A–DNA. The Raman spectroscopy represents a suitable tool for characterization of structural changes of DNA modified by the chosen lanthanides ions.

Acknowledgement

This work on this project was supported by the grant GACR P301/10/0356.

References

Duguid JG, Bloomfield VA, Bene vides JM, Thomas GJ (1995) Raman spectroscopy of DNA-Metal complexes .2. The thermal denaturation of DNA in the presence of Sr2+, Ba2+, Mg2+, Ca2+, Mn2+, Co2+, Ni2+, and Cd2+. Biophysical Journal 69(6):2623-2641

Fricker SP (2006) The therapeutic application of lanthanides. Chemical Society Reviews 35(6):524-533

Schmitt M, Popp J (2006) Raman spectroscopy at the beginning of the twenty-first century. Journal of Raman Spectroscopy 37(1-3):20-28

Vrana O, Masek V, Drazan V, Brabec V (2007) Raman spectroscopy of DNA modified by intrastrand cross-links of antitumor cisplatin. Journal of Structural Biology 159(1):1-8

J Biochem Tech (2010) 2(5):S102-S103
ISSN: 0974-2328

Study of humic substances by fluorescence spectroscopy

Sona Konecna, Nadezda Fasurova, Martina Klucakova*

Received: 25 October 2010 / Received in revised form: 13 August 2011, Accepted: 25 August 2011, Published: 25 October 2011
© Sevas Educational Society 2011

Abstract

The purpose of this study is to determine main fluorophores of soil humic substances using 2D and 3D synchronous fluorescence spectroscopy (SFS). The measured synchronous spectra were compared with standards IHSS. Differences between humic and fulvic acids as well as our and IHSS samples are discussed.

Keywords: Humic substances, synchronous fluorescence spectroscopy.

Introduction

Athough the optical properties of humic substances (HS) have been investigated by many authors, the structural basis of these properties remains unclear. HS contain conjugated olefinic, aromatic, phenolic–semiquinone–quinone structures with functional groups and chromophores, which have been utilized in their characterization by fluorescence spectroscopy. Obtained results provide strong evidence that the absorption and emission spectra must arrise from a very large number of absorbing and emitting species or states (Boyle et al. 2009; Del Vecchio and Blough 2004). Total luminiscence spectra showed that the fluorescence spectra may be used to disriminate between soil-derived and aquatic derived IHSS humic substances and between humic and fulvic acids isolated from the same source (Mobed et al. 1996). Other authors utilized fluorescence spectroscopy for determination of humification degree of soil and compost humic acids (Milori et al. 2002; Fuentes et al. 2006) or characterization of HS isolated from different soil types (Barančíková et al. 1997). The effect of preparation procedure on the structural integrity of humic acids was investigated by fluorescence spectroscopy using pyrene as a hydrophobic probe. Changes in native fluorescence as well as pseudo-micelle formation were deduced on the basis of measured spectra (Engebretson and Wandruszka 1999).

Sona Konecna, Nadezda Fasurova, Martina Klucakova*

Institute of Physical and Applied Chemistry, Faculty of Chemistry, Brno university of Technology, Purkyňova 118, 612 00 Brno

*Tel: +420 541 149 410, Fax: +420 541 211 697
E-mail: klucakova@fch.vutbr.cz

Extending the 2D SFS to 3D SFS in this work gives the total synchronous characteristics of multifluorophoric sample of HS at various possible wavelength intervals, which could help to characterize better multifluorophoric systems (as humic substances).

Materials and methods

HA and FA extracted from cambisol, cambisol leach (0,1 M sodium pyrophosphate and 0,1 M NaOH) and standard IHSS Elliot HA and FA were used in this work. Details of HA and FA extraction as well as used measurement technique are described in (Patra and Mishra 2002).

2D SFS spectra were measured within range 200 nm and 600 nm by Spectrofluorimeter AMINCO Bowman Serie 2, at fixed wavelength interval ($\Delta\lambda = 40$ nm) between excitation and emissionmonochromators. 3D SFS spectra were scaned from 200 to 600 nm with gradual increase of $\Delta\lambda$ from 30 to 150 nm with step 5 nm. This allows detailed characterization present fluorophores and resolution peaks, that we cannot see by measure 2D synchronous scan with one $\Delta\lambda$ value.

Results and Discussion

The 2D spectra in Fig. 1 showed IHSS Elliot FA and cambisol FA have the same course with maximum at emission wavelength 399 nm, which is lower wavelength than that obtained for IHSS Elliot HA and cambisol HA (468 and 507 nm, respectively). The cambisol leach had two dominant peaks at 507 nm, which signifies presence of HA and at 379 nm, which signifies presence of FA, because the leach contain both types of HS.

3D SFS spectrum of cambisol HA is obtained by plotting fluorescence intensity as a combined function of the emission wavelength and of the difference between emission and excitation wavelenght $\Delta\lambda$ [2]. Two main fluorophores groups were found in cambisol HA sample: $\lambda_{exc.}= 458$ nm (for $\lambda_{em.}= 468$ nm) and $\lambda_{exc.}= 396$ nm (for $\lambda_{em.}= 501$ nm), where the first one is identical in the emission maximum with emission maximum for 2D spectrum.

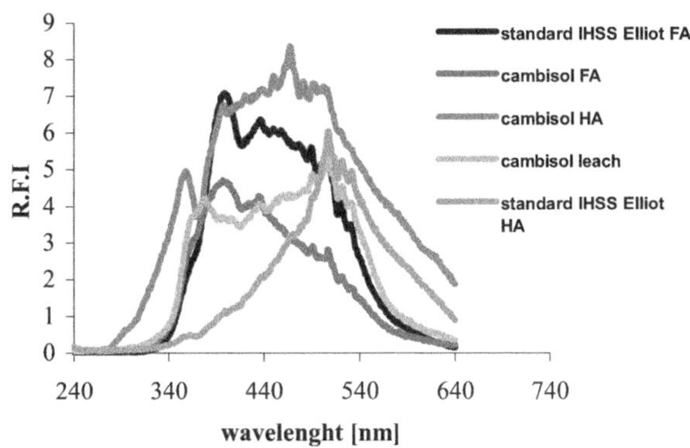

Figure 1: 2D synchronous spectrum with fixed wavelength (Δλ= 40 nm)

Conclusion

Differences in intensity and location of main peaks were observed for HA and FA. Whereas relatively simple aromatic/phenolic compounds were observed in FA spectra, extensively conjugated phenolics and quinones were identified for HA samples. Method of 3D SFS allowed characterization of main fluorophores, while method of 2D SFS is able to scan by only one fixed wavelength.

Acknowledgement

The work has been supported by Ministry of Education (project MSM 0021630501).

References

Barančíková G, Senesi N, Brunetti G (1997) Chemical and spectroscopic characterization of humic acids isolated from different Sloval soil types. Geoderma 78:251-266

Boyle ES, Guerriero N, Thiallet A et al. (2009) Optical properties of humic substances and CDOM: Relation to structure. Environ Sci Technol 43:3362-2268

Del Vecchio R, Blough NV (2004) On the origin of the optical properties of humic substances. Environ Sci Technol 38:3885-3891

Engebretson RR, von Wandruszka R (1999) Effects of humic acid purification on interactions with hydrophobic organic matter: Evidence from fluorescence behaviour. Environ Sci Technol 33:4299-4303. Environ Sci Technol 38:3885-3891

Fuentes M, Gonzáles G, García-Mina JM (2006) The usefulness of UV-visible and fluorescence spectroscopies to study the chemical nature of humic substances from soils and composts. Org Geochem 37:1949-1959

Milori D, Martin-Neto L, Bayer C et al. (2002) Humification degreeof soil humic acids determined by fluorescence spectroscopy. Soil Sci 1467:739-749

Mobed JJ, Hemmingsen SL, Autry JL et al. (1996) Fleorescence characterization of IHSS humic substances: Total luminiscence spectra with absorbance correction.

Patra D, Mishra AK (2002) Total synchronous fluorescence scan spectra of petroleum products. Anal Bioanal Chem 373:304-309. Environ Sci Technol 30:3061-3065

J Biochem Tech (2010) 2(5):S104-S105
ISSN: 0974-2328

Comparison of complexation properties of humic acids and simple organic ligands

Martina Klucakova*, Radka Bachrata, Andrea Kargerova

Received: 25 October 2010 / Received in revised form: 13 August 2011, Accepted: 25 August 2011, Published: 25 October 2011
© Sevas Educational Society 2011

Abstract

Simple organic compounds as citric acid, phthalic acid, salicylic acid, EDTA, hydroquinone and pyrocatechol were used as structural models of active sites of humic acids. Combination of high resolution ultrasound spectrometry with potentiometry, conductometry and UV/VIS spectrometry were utilized for complexation of copper (II) ions by humic acids and model compounds. Changes in the slope of measured quantities were used to find the saturation of binding. Ultrasound spectrometry showed follow changes of hydration of interacting species. The differences observed for individual model compounds show that there are active centres not only with various strength and stability of formed complexes but also with their various rigidity and ability of conformational changes.

Keywords: Humic acid, copper ions, high resolution ultrasound spectrometry, potentiometry, UV/VIS spectrometry, conductometry

Introduction

Several simple organic (hydroxy)acids as models of principal functional moieties of lignite humic acids and the copper ion as a model ion to investigate metal–humic interactions were studied by four various methods. The models were chosen for their. similarity with the active sites in the structure of humic acids and high affinity to metal ions, e.g. salicylic acid is frequently considered as the most suitable coordination site in structure of humic acids (Saar and Weber 1980). No significant differences in the complexation properties of various phenolic acids as the ligand models of humic acids towards Cu^{2+} ions were observed. Because all these acids contained the catechol group, authors postulated that the complexation process is intrinsically related with its presence (Borges et al. 2005).

Martina Klucakova*, Radka Bachrata, Andrea Kargerova

Brno University of Technology, Faculty of Chemistry, Purky ňova 118, 612 00 Brno,

*Tel: +420 541 149 410, Fax: +420 541 211 697
E-mail: klucakova@fch.vutbr.cz

Many authors reported that there are four possibilities of humic binding with metal ions: 1) chelation between carboxylic and phenolic groups, 2) chelation between two carboxylic groups, 3) complexation with carboxyl group and 4) binding with phenols or phenolic ethers (Baker and Khalili 2003; Pandey et al. 1999).

A model of fulvic acids based on randomized positioning of the functional groups and aromatic rings showed that the predominant bidentate sites are likely to be phthalate- and salicylate-type sites with a significant proportion of the aromatic carboxylic and phenolic groups not participating in the chelating sites (Murray and Linder 1984). The model was then applied to both humic and fulvic acids and several models included the nitrogen atom were added (Bryan et al. 1997). The important role of N content in the affinity of humic substances towards copper was predicated also by other authors (Gregor et al. 1989).

The method of ultrasound spectrometry is newly utilized in study of the humic complexation in this work. The method is based on observing interactions of the ultrasound wave with the studied system. The ultrasound velocity reflects the local elasticity and density of the material, which are determined by the molecular arrangements, conformation and the solvation shell. Due the high resolution the method is sensitive to, e.g., details of the molecular organization or the solvating state. Applications thus may cover the study of micellization or aggregation, ligand-ligate binding, metal ion-(bio)polymer interaction, conformation changes etc.

Materials and methods

Ultrasonic spectrometer with high resolution HR–US 102 (Ultrasonic Scientific, Ireland), was utilized for measurement of basic ultrasonic parameters at three various frequencies (5110, 8220 and 12 200 kHz). No changes of ultrasonic velocity (ΔU) on applied frequency were occurred and presented results are the average values of these three determinations. The device consists of two independent cells tempered at 25°C. Both cells were filled by the same model compound or humic acids solution (1 cm^3). Sample in cell 1 was titrated by $CuCl_2$ (1 $mmol.dm^{-3}$) up to saturation of acidic group. The conductometric and potentiometric titrations were carried out with the same solutions to investigate complexation of humic acids and model compounds in detail. The solutions of model compounds or humic acids (50 cm^3) were titrated by CuCl$_2$ (1 mmol.dm^{-3}) up to saturation of acidic group as in previous case. The combined pH/conducto-meter (Mettler Toledo S47 SevenMulti^{TM}) was used for monitoring pH values and conductivity after each addition of titrant.

Results and Discussion

In Fig 1 we can see that titration of EDTA causes gradually the increase of ultrasound velocity (ΔU) but the concentration dependence is not linear. The decrease in slope between 0,3 and 0,7 value of Cu/H (ratio between added Cu^{2+} ions and amount of acidic groups of model) corresponding with saturation of model active sites was observed. Measured pH–values confirmed that no H$^+$ ions are splitted off for Cu/H > 0,7. Titration of humic acids by CuCl$_2$ showed minimum of ΔU at Cu/H ratio slightly above 0,5. It corresponds which hypothesis that the bivalent ions are bonded by two acidic functional groups.

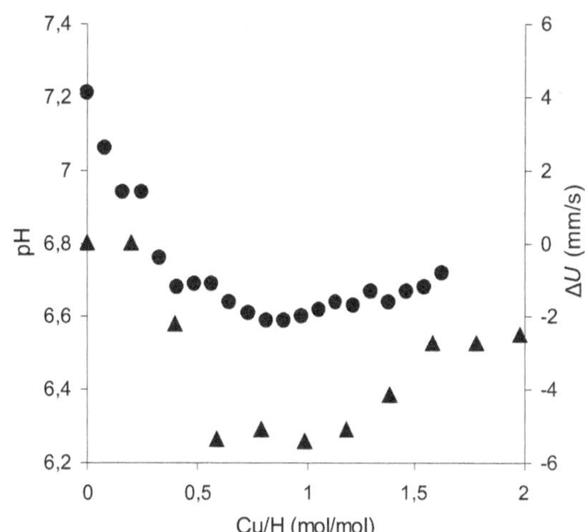

Figure 2: Comparison of results obtained by potentiometry and ultrasound spectrometry for titration of humic acids by CuCl$_2$

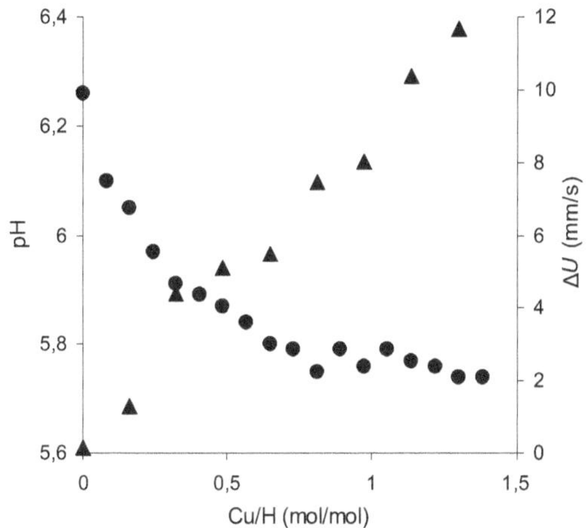

Figure 1: Comparison of results obtained by potentiometry and ultrasound spectrometry for titration of EDTA by CuCl$_2$

The ΔU-value obtained for complexation of humic acids decreases with the addition of Cu^{2+} ions and the obtained concentration dependence has a relatively marked minimum (in contrast to the used model compounds). The formation of Cu-humic complexes results in the initial decrease of ΔU. It is caused by the release of hydration water from the coordination shell of copper and atomic groups of humic acids. The compressibility of water in the hydration shells of the ligand is less than that of bulk water therefore transferring of hydration water into the bulk water increases the total compressibility of the solution, thus reducing the ΔU-value. When all available sites are occupied the curve levels off. The electrostatically neutralised humic molecules begin to form the aggregates tt higher concentrations of copper, which could cause the increase of ΔU at this stage.

Conclusion

The interaction of humic compounds with metal ions is complex problem, studied by various methods to understand better their principles and effects on their function in nature. The differences observed for individual model compounds show that there are active centers not only with various strength and stability of formed complexes but also with their various rigidity and ability of conformational changes.

Acknowledgement

This work was supported by government funding – Czech Science Foundation, project. Nr. 104/08/0990 and by the project "Centre for Materials Research at FCH BUT" No. CZ.1.05/2.1.00/01.0012 from ERDF.

References

Baker H, Khalili F (2003) Comparative study of binding strengths and thermodynamic aspects of Cu(II) and Ni(II) with humic acid by Schubert's ion-exchange method. Anal Chim Acta 497: 235-248

Borges F, Guimaraes C, Lima JL et al. (2005) Potentiometric studies on the complexation of copper(II) by phenolic acids as discrete ligand models of humic substances. Talanta 66: 670-673

Bryan ND, Robinson VJ, Livens FR et al. (1997) Metal-humic interactions: A random structural modelling approach. Geochim Cosmochim Acta 61: 805-820

Gregor JE, Powell HKJ, Town RM (1989) Evidence for aliphatic mixed mode coordination in copper(II)-fulvic acid complexes. J Soil Sci 40:661-673

Murray K, Linder PW (1984) Fulvic acids - Structure and metal binding. 2. Predominant metal binding sites. J Soil Sci 35:217-222

Pandey AK, Pandey SD, Misra V et al. (1999). Formation of soluble complexes of metals with humic acid and its environmental significance. Chem Ecol 16: 269-282

Saar RA, Weber JH (1980) Comparison of spectrofluorometry and ion-selective electrode potentiometry for determination of complexes between fulvic acid and heavy metal ions. Anal Chem 52:2095-2100

J Biochem Tech (2010) 2(5):S106-S107
ISSN: 0974-2328

Study of elemental distribution in urinary stones by Laser Ablation Inductively Coupled Plasma Mass Spectrometry (LA-ICP-MS)

Katerina Proksova, Karel Novotny, Jozef Kaiser, Michaela Galiova, Tomas Vaculovic, Viktor Kanicky*

Received: 25 October 2010 / Received in revised form: 13 August 2011, Accepted: 25 August 2011, Published: 25 October 2011
© Sevas Educational Society 2011

Abstract

Bio–mineral structures, i.e., bones, teeth or kidney stones have been found to be excellent "archives" related to nutrition, living habits and exposure to changing environmental conditions. More specifically, formation of urolithic concrements (uroliths) represents specific biomineralization of living organism. Line scans of the concrement cross–sections may provide information about the accumulation history of the elements of interest.

Analytical methods which can map the presence of different elements are e.g. Laser Induced Breakdown Spectroscopy (LIBS) and Laser Ablation Inductively Coupled Plasma Mass Spectrometry (LA–ICP–MS). Both of these methods have proven to be suitable for mapping the matrix elements, so further work is concerned with their use for the mapping of trace elements.

Keywords: LA-ICP-MS, urinary stones, distribution, major and trace elements

Introduction

One of the subjects in biomedical research is to determine the content and distribution of individual elements in human tissues. The main objective is to explain the mechanism of accumulation of selected elements in various organs, which can cause diseases and disorders of these organs.

Katerina Proksova, Karel Novotny, Michaela Galiova, Tomas Vaculovic, Viktor Kanicky*

Department of Chemistry, Faculty of Science, Masaryk University, Kotlářská 2, 611 37 Brno, Czech Republic

Jozef Kaiser

Institute of Physical Engineering, Faculty of Mechanical Engineering, Brno University of Technology, Technická 2896/2, 616 69 Brno, Czech Republic

*Tel: +420 549 494 774, Fax: +420 549 492 443
E-mail: viktork@sci.muni.cz

Creation of urolithiasis is one of most common human diseases, usually with very unpleasant and painful syndromes, as they may find difficulty in urinating, and small fragments of blood in the urine (Sperrin et al. 2002).

Urinary stones are divided into several groups according to their position within the body and also by the principal components of stones. The main constituents of urinary concretions are phosphate, oxalate and urate (Moroz et al. 2009).

In the past, for the quantitative and qualitative analysis of urinary stone composition mainly solution analysis was utilized, i.e. the stones were dissolved prior this analysis. The biggest disadvantage of this approach is that valuable information on the structure, distribution of individual elements or the composition of the nucleus cannot be obtained. These data are however especially important not only to clarify the beginning and growth of urinary stones, but also to facilitate the healing process.

Using laser ablation based methods, e.g. LA–ICP–MS; it is possible to study the distribution of elements across a variety of important lines in stones (Chaudhri et al. 2007). This method allows us to distinguish between urinary stone structures, as the core or the outer edge. Also, information about the correlation of selected major and trace elements can be obtained from the 2D or 3D maps of these elements.

Here we demonstrate that the LA-ICP-MS in fast line scan mode is sufficiently sensitive for elemental mapping most of important minor and major elements that are present in uroliths. The results obtained by LA-ICP –MS measurements were also compared with results obtained by infrared spectroscopy.

Materials and methods

The investigated samples were urinary stones surgically removed from the patient's body. After the removal the stones were rinse with water and dried. Infrared spectroscopy was performed in order to identify the major components and the type of urinary stone. For subsequent LA–ICP–MS mapping the stones were cut into half. Part of the stone was embedded in epoxy resin, and mapped. All measured samples were from the collections of Professor Petr Martinec, UGN Ostrava.

For LA the UP – 213 ablation system (New Wave Research, USA), equipped with a Nd:YAG laser emitting radiation at a wavelength of 213 nm was used. The created aerosol was introduced using a carrier gas into the inductively coupled plasma with quadrupole mass analyzer and electron photomultiplier detector Agilent 7500ce (Agilent, Japan). The ablation laser was used in line scan mode. The optimized parameters, fixed during the measurements were: integration time 0.01 s, repetition rate 5 Hz, the feed rate (speed of movement) of the sample 40 μm s^{-1}, laser power density 6.56 J cm^{-2} and beam diameter 65 μm. Set of ablation lines was directed across the whole surface of the sample cross section and individual lines were spaced at the distance of 100 μm.

The signals obtained for selected isotopes during ablation of each line scans were then stacked in 2D maps using Grams32 software.

Results and Discussion

As an example, Fig 1 shows the ^{44}Ca distribution on the cross–section of urinary stone no. 11 727. This sample is oval bladder stones, for which the contents of major components were determined using infrared spectroscopy. This is a whewellite sample, which is covered on the surface with a thin apatite layer.

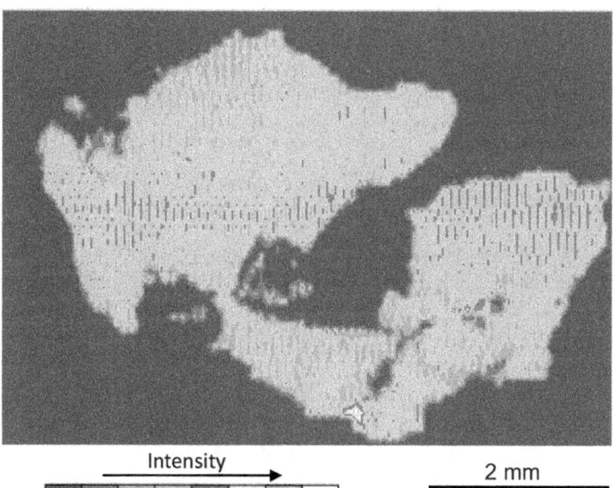

Figure 1: Map of ^{44}Ca

The following isotopes of selected elements were measured: ^{12}C, ^{23}Na, ^{24}Mg, ^{31}P, ^{39}K, ^{43}Ca, ^{44}Ca, ^{55}Mn, ^{56}Fe, ^{63}Cu, ^{66}Zn, ^{109}Ag, ^{118}Sn, ^{208}Pb. Maps were created for all isotopes and it was found that the main ingredient of this stone is ^{44}Ca.

It was also observed, that in several places of the stone cross–section the reduction in ^{44}Ca content is in balance with the increment of ^{12}C. This indicates the presence of calcium oxalate. In some places trace contents of ^{31}P were detected, which may point out the presence of apatite (calcium phosphate).

The texture and the quantitative mineralogical composition of selected calcium oxalate–based urinary stone fragments were also investigated by synchrotron radiation X–ray microtomography.

Conclusion

Maps created by LA–ICP–MS confirmed the results obtained from infrared spectroscopy. The selected urinary stone was whewellite (calcium oxalate) with traces of apatite (calcium phosphate). On the frame of the ongoing work the stone is further investigated. The presence of trace elements and their correlation with matrix elements is studied.

Acknowledgement

We acknowledge the Ministry of Education, Youth and Sports of the Czech Republic for bestowing the research projects OC09013, E08002 and MSM0021622412 and the Czech Grant Agency for the project GA203/09/1394. J.K. Acknowledges Central European Initiative (CEI) for Ceres fellowship.

References

Chaudhri MA, Watling J, Khan FA (2007) Spatial distribution of major and trace elements in bladder and kidney stones. Journal of Radio analytical and Nuclear Chemistry, 271(3):713 – 720

Moroz TN, Palchik NA, Dar'in AV (2009) Microelemental and mineral compositions of pathogenic biomineral concrements: SRXFA, X–ray powder diffraction and vibrational spectroscopy data. Nuclear instruments & methods in physics research section A – Accelerators spectrometers detectors and associated equipment, 603(1-2):141 – 143

Sperrin M, Rogers K, Lane D (2002) An investigation into the architecture and composition of a urinary calculus. Journal of materials science 13(1):7–9

J Biochem Tech (2010) 2(5):S108-S109
ISSN: 0974-2328

Spectroscopic study of selenadiazoloquinolones in alkaline media

Zuzana Barbierikova, Andrej Stasko, Michal Zalibera, Maros Bella, Viktor Milata, Vlasta Brezova*

Received: 25 October 2010 / Received in revised form: 13 August 2011, Accepted: 25 August 2011, Published: 25 October 2011
© Sevas Educational Society 2011

Abstract

Newly synthesized derivatives of 6–oxo–6,9–dihydro[1,2,5] selenadiazolo [3,4–h] quinoline variously substituted at position 7 are witnessed in solution in the N_9–deprotonated and protonated oxo tautomeric forms depending on the pH using UV/vis and NMR spectroscopy. Upon the anodic oxidation selenadiazoloquinolones produce paramagnetic species observed by EPR spectroscopy.

Keywords: UV/vis spectroscopy, NMR spectroscopy, selenadiazoloquinolones.

Introduction

4–Oxoquinolines represent a group of heterocyclic compounds widely applied in the medical care. The presence of selenium in molecules can indicate new effects, and some of synthesized selenaheterocyclic compounds demonstrated attractive biological impact. Novel 7–substituted 6–oxo–6,9–dihydro[1,2,5] selenadiazolo[3,4–h]quinoline (R = H, $COOC_2H_5$, $COOCH_3$, COOH, $COCH_3$ and CN) were synthesized as potential anticancer and antimicrobial agents. The presented work deals with the formation of radical species from selenadiazoloquinoline derivatives generated upon electrochemical oxidation in alkaline media at various pH values and monitored by *in situ* EPR spectroscopy.

Zuzana Barbierikova, Andrej Stasko, Michal Zalibera, Vlasta Brezova*

Institute of Physical Chemistry and Chemical Physics, Faculty of Chemical and Food Technology, Sl ovak University of Technology in Bratislava, Radlinského 9, SK–812 37 Bratislava, Slovak Republic

*Tel: +421 (2) 59 325 66, Fax: +421 (2) 52 496 032
E-mail: vlasta.brezova@stuba.sk

Maros Bella, Viktor Milata

Institute of Organic Chemistry, Catalysis and Petrochemistry, Faculty of Chemical and Food Technology, Slovak University of Technology in Bratislava, Radlinského 9, SK–812 37 Bratislava, Slovak Republic

Materials and methods

The electrochemical EPR experiments were performed in the alkaline aqueous solutions (0.1 M or 0.001 M NaOH) saturated with selena–diazoloquinolones. ^{1}H and ^{13}C NMR spectra were recorded with Varian VNMRS 600 MHz spectrometer equipped with the ^{13}C Enhanced Salt Tolerant Triple Resonance coldprobe in 0.1M NaOH/10% D_2O solution at 25°C and acetone as reference standard (^{1}H/^{13}C methyl 2.22/30.89 ppm). The UV/visible spectra of investigated selenadiazoloquinolones in aqueous solutions at different pH were recorded using a UV–3600 UV/Vis spectrometer (Shimadzu, Japan).

Results and Discussion

UV/vis experiments showed that two forms of selenadiazoloquinolones are present in the solution, depending on the pH. As the observed behavior cannot be explained by the

(a)

(b)

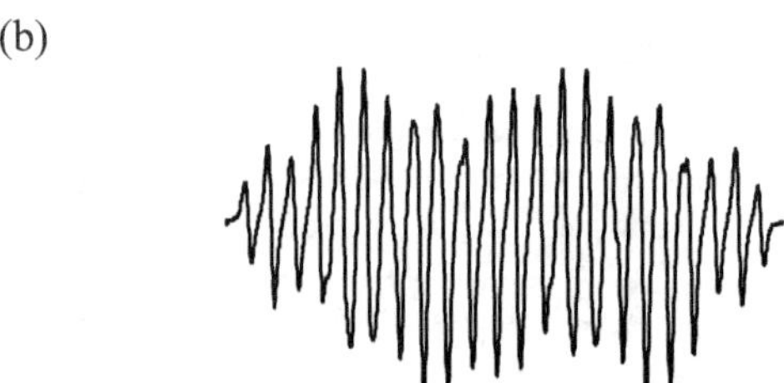

Figure 1: EPR spectra observed by the anodic oxidation of 6–oxo–6,9–dihydro[1,2,5] selenadiazolo[3,4–h]quinoline in aqueous (a) 0.1 and (b) 0.001 NaOH (magnetic field sweep width 1.0 mT).

tautomeric hydroxy/oxo exchange (as evidenced by the NMR data) the two forms can only result from the deprotonation/protonation of the acidic hydrogen at N9 of the 4–pyridone moiety. The EPR spectra measured upon the anodic oxidation of unsubstituted derivative in 0.1 and 0.001 NaOH originate from the N9–deprotonated and protonated oxo–tautomeric forms, respectively (Fig 1). However, their simulation analysis revealed no hyperfine splittings corresponding to N1, N3 and 77Se nuclei. Consequently, we assume that absence of primary paramagnetic oxidation products is caused by the alkaline conditions coupled with high polarization potential (~1.5 V), and the initial oxidation step is followed by hydrolysis of selenadiazolo moiety producing ortho-semiquinone paramagnetic species.

Conclusion

Under alkaline conditions investigated selenadiazoloquinolones exist in the oxo–tautomeric forms. EPR *in situ* electrochemical oxidation revealed the formation of paramagnetic species most probably characterized with semiquinone structure.

Acknowledgement

This work was supported by the Research and Development Agency of the Slovak Republic (contracts No. APVT–0055–07) and the Scientific Grant Agency of the Slovak Republic (Projects VEGA/1/0018/09 and 1/0225/08) and the State Program Grant 2003SP200280203.

References

Odom B, Hanneke D, D'Urso B, Gabrielse G (2006) New Measurement of the Electron Magnetic Moment Using a One-Electron Quantum Cyclotron. Physical Review Letters 97 (3):030801

J Biochem Tech (2010) 2(5):S110-S111
ISSN: 0974-2328

Comparison of complexation properties of humic acids and simple organic ligands

Martina Klucakova*, Radka Bachrata, Andrea Kargerova

Received: 25 October 2010 / Received in revised form: 13 August 2011, Accepted: 25 August 2011, Published: 25 October 2011
© Sevas Educational Society 2011

Abstract

Simple organic compounds as citric acid, phthalic acid, salicylic acid, EDTA, hydroquinone and pyrocatechol were used as structural models of active sites of humic acids. Combination of high resolution ultrasound spectrometry with potentiometry, conductometry and UV/VIS spectrometry were utilized for complexation of copper (II) ions by humic acids and model compounds. Changes in the slope of measured quantities were used to find the saturation of binding. Ultrasound spectrometry showed follow changes of hydration of interacting species. The differences observed for individual model compounds show that there are active centres not only with various strength and stability of formed complexes but also with their various rigidity and ability of conformational changes.

Keywords: Humic acid, copper ions, high resolution ultrasound, potentiometry, UV/VIS spectrometry, conductometry

Introduction

Several simple organic (hydroxy)acids as models of principal functional moieties of lignite humic acids and the copper ion as a model ion to investigate metal–humic interactions were studied by four various methods. Models were chosen mainly for their high affinity to metal ions, e.g. salicylic acid is frequently considered as the most suitable coordination site in structure of humic acids.

Materials and methods

Ultrasonic spectrometer with high resolution HR–US 102 (Ultrasonic Scientific, Ireland), was utilized for measurement of basic ultrasonic parameters at three various frequencies (5110, 8220

Martina Klucakova*, Radka Bachrata, Andrea Kargerova

Brno University of Technology, Faculty of Chemistry, Purky ňova 118, 612 00 Brno,

*Tel: +420 541 149 410, Fax: +420 541 211 697
E-mail: klucakova@fch.vutbr.cz

and 12 200 kHz). No changes of ultrasonic velocity on applied frequency were occurred and presented results are the average values of these three determinations. The device consists of two independent cells tempered at 25 °C. Both cells were filled by the same model compound or humic acids solution (1 cm^3). Sample in cell 1 was titrated by $CuCl_2$ (1 mmol dm^{-3}) up to saturation of acidic group. The conductometric and potentiometric titrations were carried out with the same solutions to investigate complexation of humic acids and model compounds in detail (Banerjee et al. 1990).

Results and Discussion

We can see that titration of EDTA causes gradually the increase of ultrasound velocity (ΔU) but the concentration dependence is not linear. The decrease in slope between 0,3 and 0,7 value of Cu/H (ratio between added Cu^{2+} ions and amount of acidic groups of model) corresponding with saturation of model active sites was observed. Measured pH–values confirmed that no H^+ ions are splitted off for Cu/H > 0.7.

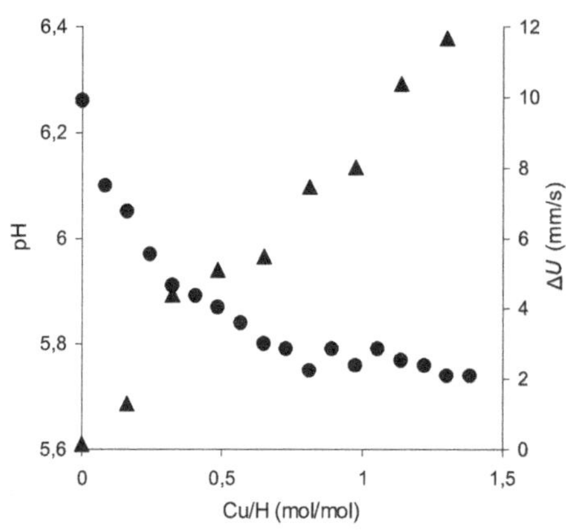

Figure 1: Comparison of results obtained by potentiometry and ultrasound spectrometry for titration of EDTA by $CuCl_2$

Titration of humic acids by $CuCl_2$ showed minimum of ΔU at Cu/H ratio slightly above 0,5. It corresponds which hypothesis that the bivalent ions are bonded by two acidic functional groups

Conclusion

The interaction of humic compounds with metal ions is complex problem, studied by various methods to understand better their principles and effects on their function in nature. The differences observed for individual model compounds show that there are active centers not only with various strength and stability of formed complexes but also with their various rigidity and ability of conformational changes.

Acknowledgement

This work was supported by government funding – Czech Science Foundation, project. Nr. 104/08/0990

References

Banerjee S, M K Gunasekaran, A K Raychaudhuri (1990) A phase-sensitive superheterodyne ultrasonic spectrometer. Meas Sci Technol 1:505

J Biochem Tech (2010) 2(5):S112-S113
ISSN: 0974-2328

Mapping of elements distribution in plant samples using LIBS

Lucie Krajcarova, Karel Novotny, Michaela Galiova, Jozef Kaiser, Viktor Kanicky, Vojtech Adam, Re ne Kizek*, Petr Babula

Received: 25 October 2010 / Received in revised form: 13 August 2011, Accepted: 25 August 2011, Published: 25 October 2011
© Sevas Educational Society 2011

Abstract

This work is focused on application of LIBS technique for the study of plant samples. The elemental mapping on cross section throughout the annual terminal twigs of *Picea abies* was performed using double pulse Laser Induced Breakdown Spectroscopy (DP LIBS). 2D maps were created, where distribution of Cu and Ca in the plant tissue was observed. After mineralization of twig parts originated in the vicinity of the implementation of cross section, ICP–MS analysis was used for determination of total content of investigated elements.

Keywords: Laser Induced Breakdown Spectroscopy (LIBS), *Picea abies,* heavy metals

Introduction

Heavy metals contained in soil are of great interest due to toxic metal accumulation in plants. Heavy metals can get to higher parts

Lucie Krajcarova, Karel Novotny, Michaela Galiova, Viktor Kanicky

Department of Chemistry, Masaryk University, Brno, Czech Republic

Jozef Kaiser

Institute of Physical Eengineering, University of Technology, Brno, Czech Republic

Vojtech Adam, Rene Kizek

Department of Chemistry and Biochemistry, Mendel University of Agriculture and Forestry, Brno, Czech Republic

*Tel: +420 545 133 350, Fax: +420 545 212 044
E-mail: kizek@sci.muni.cz

Petr Babula

Department of Natural Drugs, University of Veterinary and Pharmaceutical Sciences, Brno, Czech Republic

of food chain, even to human. There are some possibilities to remediate these soils. One of them is phytoremediation, which uses plants to accumulate heavy metals from soil. Monitoring of these metals in plant organs and tissues is important both, for protecting the human health and for checking the phytoremediation processes. In previous works, LIBS technique was successfully used for monitoring the distribution of elements in different parts of plants (Galiova et al. 2008; Galiova et al. 2007; Kaiser et al. 2009, Kaiser et al. 2007). Copper is an important trace element but its excess can have negative influence for plants, e.g. some nutrition elements intake can be affected. High Cu amount can cause specific toxicity symptoms also in human body.

Materials and methods

Annual terminal twigs of spruce (*Picea abies*) (cca 15 cm long) were cultivated in $CuCl_2$ solution at final concentrations of 1, 5, 10 and 50 mM; and in distilled water for a control samples. After 2, 4, 8, 16 and 24 hours cross sections in five places (A–E) with distance 2.5 cm from each other on twigs were made (A in the apex to E above solution surface). These slices were measured with DP LIBS in orthogonal reheating mode. Raster scanning with 150 μm spatial resolution was set. From LIBS data, 2D maps were created. Some twig parts originated in the vicinity of the implementation of cross sections were mineralized and subsequently analyzed by ICP–MS. The results give quantitative information about overall concentration of elements of interest.

Results and discussion

In the maps of Cu and Ca distribution, (Fig. 1) the transport of these ions throughout the plant via vascular bundles is observed. With increasing Cu^{2+} concentration, these ions are distributed also into surrounding tissues and less into the core.

ICP–MS showed indistinct trends of Ca. Nevertheless, visible trends of Cu were observed. The amount of Cu increases with increasing $CuCl_2$ concentration and duration of cultivation. With increasing distance from solution surface (E   A) the Cu amount decreases.

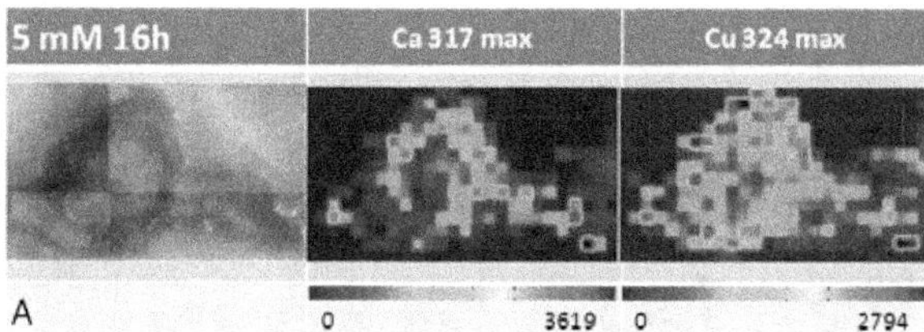

Figure 1: Maps of Cu and Ca distribution on twig slices of *Picea abies* – cultivated in 5 mM CuCl$_2$ 16 hours.

Conclusion

Maps of Cu and Ca distribution in twigs slices were made in this experiment. Results show that it is possible to use LIBS technique for observation of element distribution in plant tissues on twig slices. Data about the overall concentration of Cu and Ca in terminal twigs of *Picea abies* after different treatments were obtained by ICP–MS analysis. On the frame of ongoing work quantification of LIBS maps will be considered.

Acknowledgement

We acknowledge the Ministry of Education, Youth and Sports of the Czech Republic for research projects MSM 0021630508, MSM 0021622412, OC09013 and ME09015, ME08002.

References

Galiova M, Kaiser J, Novotny K, Novotny J, Vaculovic T et al. (2008) Investigation of heavy-metal accumulation in selected plant samples using laser induced breakdown spectroscopy and laser ablation inductively coupled plasma mass spectrometry. Applied Physics a-Materials Science & Processing 93(4):917-922

Galiova M, Kaiser J, Novotny K, Samek O, Reale L, et al. (2007) Utilization of laser induced breakdown spectroscopy for investigation of the metal accumulation in vegetal tissues. Spectrochimica Acta Part B-Atomic Spectroscopy 62(12):1597-1605

Kaiser J, Galiova M, Novotny K, Cervenka R, Reale L, et al. (2009) Mapping of lead, magnesium and copper accumulation in plant tissues by laser-induced breakdown spectroscopy and laser-ablation inductively coupled plasma mass spectrometry. Spectrochimica Acta Part B-Atomic Spectroscopy 64(1):67-73

Kaiser J, Galiová M, Novotný K, Reale L, Stejskal K et al. (2007) Utilization of the Laser Induced Plasma Spectroscopy for monitoring of the metal accumulation in plant tissues with high spatial resolution. Modern Research and Educational Topics in Microscopy 434–441

J Biochem Tech (2010) 2(5):S114-S115
ISSN: 0974-2328

Lanthanide luminescence

Pavla Pekarkova, Zdenek Spíchal, Marek Necas and Petr Taborsky*

Received: 25 October 2010 / Received in revised form: 13 August 2011, Accepted: 25 August 2011, Published: 25 October 2011
© Sevas Educational Society 2011

Abstract

Luminescent properties of trivalent lanthanides are extensively studied because of their use not only in analytical chemistry and biomedicine (Caravan et al., 1999; Selvin, 2002). In this contribution, luminescence properties of selected lanthanide complexes will be discussed.

Keywords: luminescence, lanthanides, lanthanide complexes

Introduction

Several trivalent lanthanide ions may provide noteworthy luminescent properties especially when located in coordination complexes. These include extraordinarily long luminescence life–times reaching hundreds of microseconds, Stokes shift much greater than in case of common organic fluorophores, and tight emission peaks matching the selected ion energy transitions (Choppin and Peterman 1998). An antenna effect, which allows the ligand to act as a light energy collector, is said to be responsible for such spectroscopic behavior (Selvin 2002).

Generally, the energy absorbed by the ligand may be consumed by various de–excitation ways involving irradiative processes, ligand fluorescence or phosphorescence; or it may be used for excitation and luminescence of the central ion. Thus the luminescence intensity and life–time are enormously influenced by a central ion surrounding (Bunzli and Piguet 2005).

The lanthanide complexes have been intensively studied in past few years, particularly for their use as relaxation (Gd^{3+}) reagents in magnetic resonance imaging in a medicine field of knowledge, or as labeling (Tb^{3+}, Eu^{3+}) reagents for time–resolved luminescence spectroscopy and related techniques (Caravan et al. 1999; Selvin 2002).

Pavla Pekarkova, Zdenek Spichal, Marek Necas, Petr Taborsky*

Department of Chemistry, Faculty of Science, Masaryk University, Kotlářská 2, 61137 Brno

*Tel: +420 549 495 935, Fax: +420 541 211 214
E-mail: 13423@mail.muni.cz

Materials and methods

All luminescence measurements were performed on Luminescence spectrometer AMINCO–Bowman Series 2 in quartz cuvette with optical length 10 mm at a temperature of 25°C. Data were evaluated in AB2, Microsoft Office Excel and OriginPro.

Results and Discussion

Lanthanide complexes containing various ligands provided different luminescence intensities and life–times, which was related to the first coordination sphere of the lanthanide ion. This enabled to determine a number of water molecules enclosing the central ion, which has shown to be an important issue in biomedicine, since non–complexed lanthanides are toxic (Barthelemy and Choppin 1989).

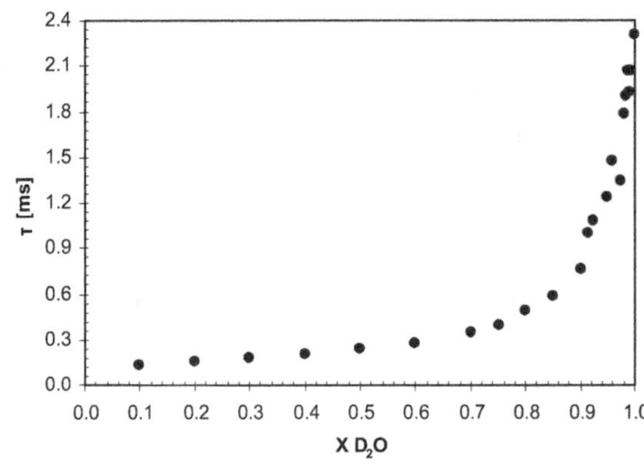

Figure 1: The luminescence life–time of Eu3+ ions as a function of D2O/H2O molar ratio

Luminescence quenching experiments in various permittivity media have been performed resulting in a strong dependence of luminescence on the presence of –XH oscillators (–OH, –NH, –CH, H_2O) in the closest Ln^{3+} ambience (Choppin and Peterman, 1998). Moreover, the quenching ability of the oscillators has been studied under different D_2O/H_2O ratios. The luminescence life–time of Eu^{3+} ions in the presence of D_2O molecules has been significantly longer than in the presence of H_2O molecules (Fig 1), which makes it suitable for D_2O purity verification or examination of water traces in other organic solvents.

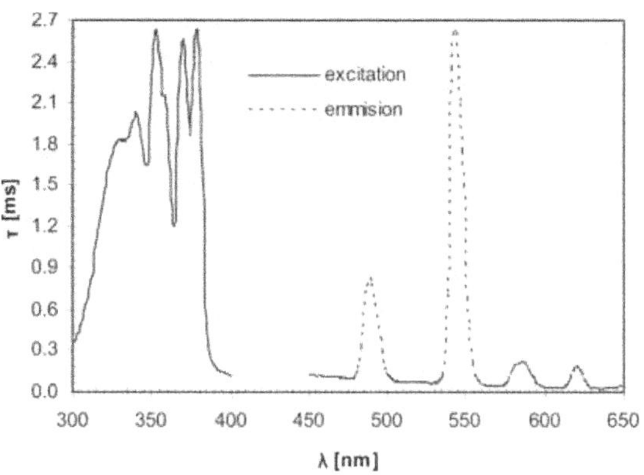

Figure 2: Excitation and emision spectrum of Tb(NO3)3 with DPPHO2 as a ligand

Finally, luminescence characterization of newly synthesized coordination compounds with bis(diphenylphosphino)alkane dioxide ligands, alkane being ethane, butane or hexane and lanthanide salt being $Tb(NO_3)_3$, $Dy(NO_3)_3$, $Sm(NO_3)_3$ and $Eu(NO_3)_3$, has been performed. Preliminary results have shown differences in luminescence lifetimes (luminescence lifetime correlated to the alkyl chain length) and emission spectra (Fig 2) of different complexes. This variation is assumed to be dependent on the length of alkane chain (Table 1).

Table 1: Luminescence properties of the studied Ln^{3+} coordination polymers

Salt	Ligand	Excitation/ emission (nm)	Life-time (ms)	Reference
$Tb(NO_3)_3$	$DPPEO_2$	343/544	1.356	This work
$Tb(NO_3)_3$	$DPPBO_2$	378/544	1.439	This work
$Tb(NO_3)_3$	$DPPHO_2$	378/543	1.539	This work
$Dy(NO_3)_3$	$DPPEO_2$	389/573	0.083	This work
$Dy(NO_3)_3$	$DPPBO_2$	453/573	0.077	This work
$Dy(NO_3)_3$	$DPPHO_2$	454/573	0.089	This work
$Sm(NO_3)_3$	$DPPEO_2$	404/597	0.075	This work
$Sm(NO_3)_3$	$DPPBO_2$	404/597	0.089	This work
$Sm(NO_3)_3$	$DPPHO_2$	404/597	0.103	This work
$Eu(NO_3)_3$	$DPPEO_2$	395/618	1.426	Pekarkova, 2011
$Eu(NO_3)_3$	$DPPBO_2$	395/618	1.566	Pekarkova, 2011
$Eu(NO_3)_3$	$DPPHO_2$	395/618	1.598	Pekarkova, 2011

Conclusion

Luminescence properties of selected lanthanide complexes have been characterized for future deployment in several fields of knowledge, such as biomedicine, quality assessment or material synthesis and development. The lanthanide complexes sphere of action is far–reaching and should be enriched by further applications in addition to fundamental research.

Acknowledgement

We gratefully acknowledge the financial support of the Czech Science Foundation (203/08/1111) and the Ministry of Education, Youth and Sports of the Czech Republic (LC Basic Research Center; LC 06035).

References

Barthelemy PP, Choppin GR (1989) Luminescence study of complexation of europium and dicarboxylic-acids. Inorganic Chemistry 28(17):3354-3357

Bunzli JCG, Piguet C (2005) Taking advantage of luminescent lanthanide ions. Chemical Society Reviews 34(12):1048-1077.

Caravan P, Ellison JJ, McMurry TJ, Lauffer RB (1999) Gadolinium(III) chelates as MRI contrast agents: Structure, dynamics, and applications. Chemical Reviews 99(9):2293-2352

Choppin GR, Peterman DR (1998) Applications of lanthanide luminescence spectroscopy to solution studies of coordination chemistry. Coordination Chemistry Reviews 174:283-299

Pekarkova P, Spichal Z, Taborsky P, Necas M (2011) Luminescent properties of europium complexes with bis(diphenylphosphino)alkane dioxides. Luminescence: The Journal of Biological and Chemical Luminescence DOI: 10.1002/bio.1291

Selvin PR (2002) Principles and biophysical applications of lanthanide-based probes. Annual Review of Biophysics and Biomolecular Structure 31:275-302

J Biochem Tech (2010) 2(5):S116-S117
ISSN: 0974-2328

Recent development of double pulse laser induced breakdown spectroscopy (DP-LIBS) setup

David Prochazka, Jozef Kaiser, Karel Novotny, Michaela Galiova, Radomir Malina*

Received: 25 October 2010 / Received in revised form: 13 August 2011, Accepted: 25 August 2011, Published: 25 October 2011
© Sevas Educational Society 2011

Abstract

Single pulse (SP) LIBS setup was modified to DP setup to achieve more accurate analytical sensitivity and spatial resolution. All parameters, like interpulse delay, acquisition delay or energy of ablation and excitation laser pulses were optimized.

Keywords: DP LIBS, SP LIBS, heavy metals

Introduction

Laser-Induced Breakdown Spectroscopy (LIBS) is a spectrochemical method that can be applied to determine elemental composition of sample using radiation of Laser-Induced Plasma (LIP). The plasma plume is created focusing the ablation laser emission. Nowadays, the mostly used LIBS lasers are nanosecond-pulsed solid state lasers, such as Q-switched Nd:YAG. The basic LIBS setup consists of one (ablation) laser, optical systems for focusing of the laser light and collection of LIP emission, respectively. The spectral analysis is realized by a spectrograph coupled usually to ICCD camera in order to carry out temporally resolved spectroscopic measurements (Miziolek et al. 2006).

Using two lasers instead of one, in so called double-pulse (DP) configuration, allows to control in some extent the LIP plasma parameters (Colao et al. 2002). As a consequence, better signal-to-noise ratio or smaller ablation crater diameters can be achieved.

David Prochazka, Jozef Kaiser, , Michaela Galiova, Radomir Malina*

Institute of Physical Engineering, Faculty of Mechanical Engineering, Technical University of Brno, Technická 2, 616 69 Brno, Czech Republic

Karel Novotny

Department of Chemistry, Faculty of Science, Masaryk University, Kotlářská 2, 616 00 Brno, Czech Republic

*Tel: +420 541 142 766, Fax: +420 541 142 842
E-mail: malina@fme.vutbr.cz

Orthogonal DP LIBS setup was realized by incorporating an another Nd:YAG laser (Solar LX-318, Belarus) to our SP LIBS setup described in details elsewhere (Kaiser et al. 2009). The DP LIP was studied at different interpulse delay times and laser pulse energies. The goal of these measurements was to obtain strong emission lines together with the smallest ablation craters.

Materials and methods

The emission lines of selected heavy metals (Pb, Cd) were observed in dried plant leaves, used also in our earlier works (Kaiser et al. 2009). In this study the acquisition delay, interpulse delay and first (ablation) and second (re-heating) laser energies were optimized.

Results and Discussion

It was shown that the emission intensities of atomic and ionic lines are strongly enhanced and the ablation crater size is reduced in the case of DP LIBS. As an example, in Table 1, a comparison of ablation craters for the same spectral line intensities from the SP and DP LIBS is shown. In order to reach the same spectral line intensities for SP method, the ablation crater size had to be enhanced approx. twice.

Table 1: Sizes of ablation craters created by the SP LIBS and DP LIBS (same spectral line intensity)

Setup	Diameter/μm
SP LIBS	100
DP LIBS	60

In Fig 1, the comparison of the spectral line intensities for SP and DP LIBS are compared. The energy E of SP LIBS is the same like the energy of ablation pulse E_1 (DP LIBS).

Conclusion

The parametric optimisation process resulted that both the DP LIPS spatial resolution and analytical sensitivity increase rapidly in

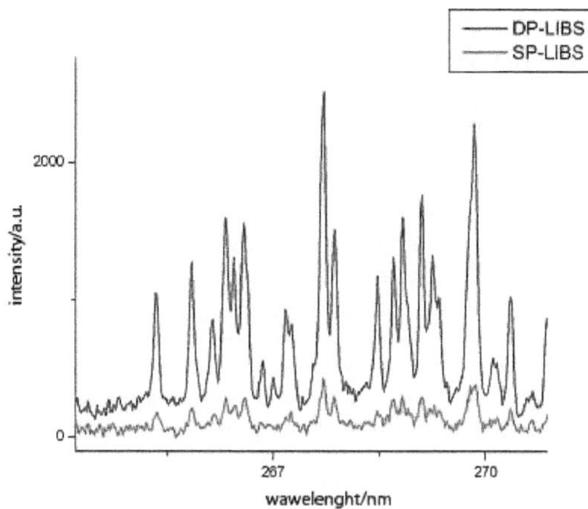

Figure 1: Comparison of line intensities for DP LIBS (blue line) and SP LIBS (red line)

comparison to SP LIBS. As a consequence, this approach can be successfully utilized for both, reduce the ablation crater size and so achieve higher spatial resolution in elemental mapping and simultaneously enhance the signal to noise ratio and achieve lower detection limits.

Acknowledgement

The authors acknowledge the Ministry of Education, Youth and Sports of the Czech Republic for research projects OC09013, ME09015, ME10061 and MSM 0021622412.

References

Miziolek A, Palleschi V, Schecter I (2006) Laser-Induced Breakdown Spectroscopy (LIBS) Fundamentals and Applications. Cambridge University Press, 638

Colao F, Lazic V, Fantoni R, Pershin S (2002) A comparison of single and double pulse laser–induced breakdown spectroscopy of aluminum samples, Spectrochimica Acta Part B: Atomic Spectroscopy, 57(7):1167–1179

Kaiser J, Galiova M, Novotny K et al (2009) Mapping of lead, magnesium and copper accumulation in plant tissues by laser-induced breakdown spectroscopy and laser-ablation inductively coupled plasma mass spectrometry. Spectrochimica Acta Part B: Atomic Spectroscopy, 64(1):67-73

J Biochem Tech (2010) 2(5):S118-S119
ISSN: 0974-2328

Study of changes of chlorophyll fluorescence in lime exposed to abiotic stress

Jiří Sochor, Petr Salas, Petr Babula, René Kizek*

Received: 25 October 2010 / Received in revised form: 13 August 2011, Accepted: 25 August 2011, Published: 25 October 2011
© Sevas Educational Society 2011

Abstract

In contribution described investigation was focused on possible utilization of additive soil substances Agrisorb, mineral zeolite and lignite in degraded soils. Photosynthetic reaction of lime - Tillia platiphyllos Scop reaction to actual soil and climatic conditions was investigated. Aim of experiment consisted on determination of effect of additive soil substances on photosynthetic processes of experimental plant.

Keywords: *Tillia platiphyllos* Scop, Chlorophyll fluorescence, soil conditions

Introduction

Global change of climate and its possible impacts are presently noticed to be of the most important fulminations of worldwide importance (Intergovernmental Panel on Climate Change). Supposed growth of global temperature will lead to changes in distribution of rainfalls and cloudiness, which will cause significant changes in Earth's ecosystems (Stine 2009).

Jiří Sochor, Petr Salas

Department of Breeding and Propagation of Horticultural Plants, Faculty of Horticulture, Mendelu, Valticka 337, 691 44 Lednice, CZ

Petr Babula

Department of Natural Drugs, Faculty of Pharmacy, University of Veterinary and Pharmaceutical Sciences, Palackeho 1-3, CZ-612 42 Brno, CZ

Rene Kizek*

Department of Chemistry and Biochemistry, Faculty of Agronomy, Mendel University in Brno, Zemedelská 1, 613 00 Brno, CZ

*Tel: +420 545 133 350, Fax: +420 545 212 044
E-mail: kizek@sci.muni.cz

In connection with global changes of climate, expanding of arid and semiarid regions can be expected. As consequence of these changes, question, whether presently used agricultural plants will be able to adapt to these extreme soil and climatic conditions, rises. In semiarid and extremely dry ecosystems of moderate climatic zone, stress to seasonal factors limiting physiological efficiency is commonly present. Reduced water availability, which interferes with other factors of dry conditions, presently represents important limitation fight against globally progressing desertification (Mercado 2009).

Additive soil substances (PPL) represent wide group of substrates, which are coming in on focus of interest also in area of recultivation. They present one of the possibilities of reduction of stress conditions for plants and improvement of chemical, physical, and biological properties of soils (Salaš 1996).

Materials and methods

With respect to aim of solving of problem, very extreme locality of agriculturally used countryside of Hodonin region, in very drying area situated on blowing sands (sthenic soil, granularity – class sand, low retention water capacity, extremely high aeration, exchange soil reaction extremely acid, pH/KCl Ah of soil horizon les than 4.5). On experimental area, model recultivation of problem locality was simulated.

For experiment itself, three PPLs were used – synthetic hydro-absorbent – preparate Agrisorb (in applied dose 20g • m^{-2}), natural untreated crushed lignite (applied dose 500g • m^{-2}) and natural mineral zeolite, fraction 2.5-5 mm (applied dose 3l • m^{-2}). As biological material, two-year-old sapling of *Tillia platiphyllos* Scop., which were cultivated on experimental area, were used. They were planted on 1.11.2008. Photosynthetic efficiency of plants was obtained by measuring of chlorophyll fluorescence. For measurement, Chlorophyll Fluorometer OS–30 (Fv/Fm) was used. Measurement itself was carried out from 1. 6. 2009 to 30. 9. 2009 in 10-day period for completely vegetative period.

Results and discussion

Main function of hydro-absorbents consists in water binding in case of its surplus (rainfalls, watering) with subsequent progressive releasing to plants. Agrisorb is organic polymeric compound, which is able of binding of water into its structure and releasing during vegetative season to roots of plants. Agrosorb is able of creeping and forming of stable gel. Gel established from 1 g of Agrosorb can bond as much as 300g of water (Salaš 1996).

Lignite is able to absorb high amount of water, in mined state contains 50% at least. This ability is reversible during processes of drying and hydratation. Natural mineral zeolite is additive soil substance of volcanic origin (tetragonal sodium aluminosilicate), which contains approximately 70 % of silicon oxide. Zeolite is highly porous with ability of water and ions absorption with subsequent releasing based on ions exchange (Salaš 1996).

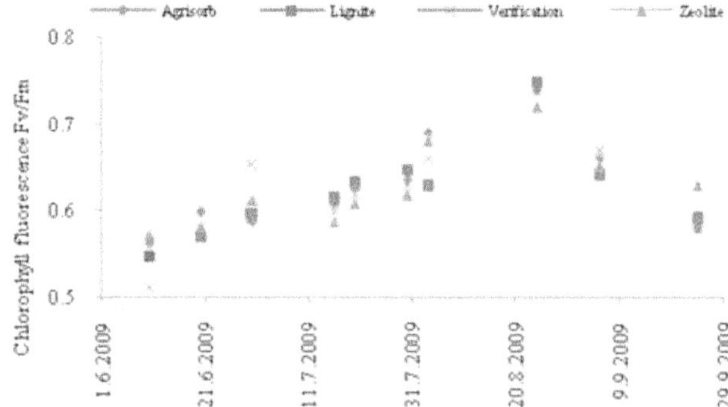

Figure 1: Comparison of values of chlorophyll fluorescence in lime determined during vegetative period on dependence of three PPLs and control area.

In the case of exposition of plant or its photosynthetic system to stress (biotic as well as abiotic), negative influence of photosystem II is recorded. Rate of variable and maximal fluorescence (Fv/Fm) is universal indicator of inhibition of photosystem II function or damage of its reaction centres. Principle of this method consists in simulation of electron excitation with different kinetics of electron exposition (Kyparissis et al. 1995).

Plants cultivated in soils containing additive soil substances demonstrated higher rate of values of variable and maximal fluorescence. Average content of chlorophyll flouorescence (ratio of variable and maximal chlorophyll fluorescence) was 0.64 for lignite, 0.61, for Agrisorb, 0.60 for zeolite and 0.58 for control variante.

Conclusion

Laboratory techniques for determination of chlorophyll fluorescence are time-consuming and destructive. Sample can be only once analysed and changes in chlorophyll fluorescence cannot be analysed during vegetative period or experiment. Described principle of determination of chlorophyll fluorescence values is based on rapid and non-destructive measurement. Due to this fact, method chosen by us proved to be the most suitable for our experiment and for measurement directly terrain.

Acknowledgement

The work has been supported by IGA 9/2010/591 a 2B08020 NPV II.

References

Stine AR (2009) Changes in the phase of the annual cycle of surface temperature. Nature, 457(7228):435

Mercado L. (2009) Impact of changes in diffuse radiation on the global land carbon sink. Nature, 458(7241):1014

Salaš P (1996) Hydroabsorbenty - látky zadržující vodu v půdě, Zahradnické produkce a lesnictví., vyd. Lednice : [s.n.],. s. 6.

Kyparissis A, Petropolou, Y, Mane tas Y (1995) Summer survival of leaves in a soft-leaved shrub (Phlomis fruticosa L. Labiatae) under Mediterranean field conditions: avoidance of photoinhibitory damage through decreased chlorophyll content. J Exp Bot 46:1825–1831

J Biochem Tech (2010) 2(5):S120-S121
ISSN: 0974-2328

Applications of MS-MALDI–TOF for quick identification of microorganisms

Petr Stursa, Michal Strejcek, Petra Junkova, Tomas Macek, Katerina Demnerova, Martina Mackova*

Received: 25 October 2010 / Received in revised form: 13 August 2011, Accepted: 25 August 2011, Published: 25 October 2011
© Sevas Educational Society 2011

Abstract

From the beginning of microbiology scientists have discussed the issue of how to easily, quickly but also accurately identify unknown microorganisms. Today there is a number of precise but time-consuming and costly molecular techniques available. The MS MALDI–TOF–based method offers a suitable alternative for currently used methods. This method allows comparing different isolates according to the characteristic profile of ribosomal proteins and selecting those which are identical, or it enables accurate direct identification of the samples using a commercial database. Measurements that we did showed that the identification accuracy by MALDI–TOF mass spectrometry is comparable to the methods of 16S rRNA gene analysis, which is used as a standard method for identification of microorganisms. The disadvantage of using commercial databases for identification is a small scale of such database. It is possible to say that method of MALDI–TOF mass spectrometry offers a fast and suitable alternative for the identification of microorganisms.

Keywords: MALDI-TOF MS, bacteria, 16S rRNA gene

Introduction

Identification of unknown microorganisms is one of the main goals, which microbiology has dealt with from its inception. There are many methods that are time consuming and technically demanding. The method of using whole cells with MALDI–TOF MS is a suitable alternative to currently used methods because it is fast and simple (Strupat et al. 1991).

Petr Stursa, Michal Strejcek, Petra Junkova, Tomas Macek, Katerina Demnerova, Martina Mackova*

Dept. of Biochemistry and Microbiology, Faculty of Food and Biochemical Technology, ICT Pra gue, Technicka 3, 166 28 Prague 6, CZ

*Tel: +420 220 445 139, Fax: +420 224 355 167
E-mail: martina.mackova@vscht.cz

Materials and methods

For the identification of microorganisms, we used the "whole cell" methodology. This method involves the growth of microorganisms on agar Luria-Bertani medium (Difco, USA) from which one colony is isolated. It is then spread on the target plate and covered with layer of 3,5-dimethoxy-4-hydroxycinnamic acid in mixture of acetonitril and 5% trifluoroacetic acid 70:30 (v/v). Crystallized samples are analyzed with mass spectrometer Biflex IV (Bruker Daltonics Inc., Billerica, USA) in linear positive ion mode with constant potential of 19 kV and pressure lower than 9.10^{-5} Pa. Each spectrum is the sum of the ions from 200 laser shots coming from different regions of the same well. The spectrum is analyzed in the range from m/z 2000-20000. The spectra are evaluated visually as well as using program BioTyper (Bruker Daltonics Inc., Billerica, USA) by creating correlation composition index.

Results and Discussion

In the measurement, we compared accuracy of the methods for the identification of microorganisms.

Table 1: Identification of microorganisms

MALDI-TOF	16SrRNA
Rhodoccocus opacus	*Rhodoccocus opacus*
Rhodoccocus opacus	*Rhodococcus sp.*
Rhodococcus sp.	*Unknown*
Rhodococcus sp.	*Rhodococcus sp.*
Rhodococcus sp.	*Rhodoccocus opacus*
Rhodococcus sp.	*Rhodoccocus opacus*
Rhodoccocus opacus	*Rhodoccocus opacus*
Achromobacter xylosooxidans	*Achromobacter sp.*
Rhodoccocus opacus	*Rhodoccocus opacus*
Rhodoccocus opacus	*Rhodoccocus opacus*
Rhodococcus sp.	*Rhodococcus sp.*
Rhodococcus sp.	*Rhodoccocus opacus*
Rhodococcus sp.	*Unknown*
Rhodococcus sp.	*Rhodococcus sp.*
Rhodococcus sp.	*Unknown*

We compared MALDI–TOF MS and analysis of 16S rRNA genes. The results are shown in the Table 1. They show that MALDI–TOF methodology is satisfactory for the identification on the species level. In cases, where identification of the isolates by analysis of 16S rRNA genes failed due to low concentration of DNA in the sample, identification by MS MALDI–TOF was successful.

Conclusion

MALDI–TOF MS is suitable to identify unknown microorganisms and the current results also show that the identification accuracy is comparable to the classically used method for analysis of 16S rRNA genes. The disadvantage of this method is the high cost of equipment and database.

Acknowledgement

The work has been supported by the MSM 6046137305, MSMT No.21/2011, MSMT 2B06156

References

Strupat K, Karas M, Hillenkamp F (1991) 2,5-Dihidroxybenzoic acid: a new matrix for laser desorption-ionization mass spectrometry. Int J Mass Spectrom Ion Processes 72(111): 89–102

J Biochem Tech (2010) 2(5): x-xli
ISSN: 0974-2328

Detailed specifications of electro-biochemical equipments

Electro-biochemical laboratory involves the combination of bio-equipments and electro-chemical equipments. Collaboration with biology or life science or electronics/electrical or allied engineering departments is required before procurement of any equipment. This section will give a brief overview of the facilities required for establishing an electro biochemical laboratory. The list of equipments, model numbers and the company names are taken from the materials and methods section of all manuscripts. The detailed specifications of all equipments are retrieved from the company website. The introductory text of the equipments is taken from Wikipedia.org, Wikimedia Foundation, USA.

Total list of equipments

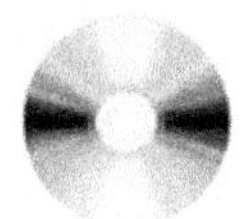

The media resources contains complete details of all equipments. It also contains information about various types of equipments required for efficient execution of electro-biochemical research projects. The software "SIMUL v5.0" can be installed directly from the compact disc. We do believe that the above list of equipments and the information present in the media resouces will help scientists in setting up electro-biochemical center for excellence (Check 'PREFACE' for downloading media resources via internet).

AMINCO-BOWMAN SPECTROFLUOROMETERS

The spectrofluorometer is an instrument which takes advantage of fluorescent properties of some compounds in order to provide information regarding their concentration and chemical environment in a sample. A certain excitation wavelength is selected, and the emission is observed either at a single wavelength or a scan is performed to record the intensity versus wavelength also called an emission spectra. Historically, in the 1950s, the National Institute of Health's Dr. Robert Bowman developed a sensitive instrument-called the spectrophotofluorometer, or "SPF"-that allowed scientists to use fluorescence as a way to identify and measure tiny amounts of substances in the body. This scientific breakthrough, invented almost half a century ago, is still used today in AIDS research and the Human Genome Project.

Spectrofluorometer specifications

System Specifications of AMINCO-Bowman Series 2 (AB2) Spectrofluorometer (FA-355, Thermo Fisher Scientific products, USA)

Light Source	Xenon flash lamp and/or 150 W continuous wave lamp (depending upon configuration – both are mountable)
Wavelength Range	220 – 850 nm
Sensitivity	CW lamp min. 2000:1 rms* Flash lamp min. 1000:1 rms**
Scan range	3 - 6000 nm/min
Slew rate	12000 nm/min
Monochromator step size	0.2 nm minimum
Wavelength accuracy	+/- 0.5 nm
Wavelength repeatability	+/- 0.25 nm
Optical Bandpass	computer control from 1 – 16 nm
Footprint	17.25" D x 31" W x 9" H (43.8 x 78.7 x 22.9 cm)
Weight (instrument only)	112 lbs (50.9 kg)
Wavelength emission and excitation correction range	200 – 600 nm
Output signals	Buffered PMT for external monitoring
PMT security	Sample Lid interlock
System compatibility	Windows 2000, NT, XP

* Signal-to-noise ratio on Raman band of water with excitation at 350 nm, emission at 397 nm and 4 nm excitation and emission slits. The emission monochromator step size is 0.2 nm and the scan rate is 0.2 nm/second. Corresponds to a minimum p-p SNR of 900:1.
** Signal-to-noise ratio on Raman band of water with excitation at 350 nm, emission at 397 nm and 8 nm excitation and emission slits. The emission monochromator step size is 1.0 nm and the scan rate is 0.2 nm/second. Corresponds to a minimum p-p SNR of 500:1.

The AMINCO-Bowman Series 2 spectrofluorometer includes a thermostattable single cuvette holder, T-Optics-configurable sample compartment, a PMT for detection, and Windows-based operating software for instrument and accessory control and data processing. Versions of the AB2 are available for 100, 120, 220 or 240 VAC, 50/60 Hz. A pre-configured and tested personal computer, monitor, and printer are recommended to complete your system. The large sample compartment and comprehensive Windows®-based software of the AB2 provide accurate and reliable measurements for busy life science laboratories. The beam geometry allows for measurements with microcuvettes (minimum volume of 125 microliters). Glan type prism polarizers provide polarization/anisotropy measurements in the UV range. Polarization measurements are made easy with the L-format and/or the optional T-format. These features allow researchers to study protein folding, DNA-protein binding, hybridization, protein-ligand binding, as well as enzyme assays and intracellular ion concentrations.

Ref: http://www2.hu-berlin.de/biologie/molbp/new/equipment/Info_AB2.pdf

Vendors (Spectrofluorometer)

Thermo Fisher Scientific Inc
1400 Northpoint Parkway,
Suite 50 West Palm Beach,FL 33407
USA
Ph: +1 800 532 4752
Email: info@thermoscientific.com

Horiba Ltd
Mr Makoto Hashimoto
2 Miyanohigashi, Kisshoin
Minami-Ku 601-8510 KYOTO Japan
Ph: +81 75 313 8121, Fax: +81 75 321 5725
Email: makoto.hashimoto@horiba.com

PerkinElmer Inc
940 Winter Street
Waltham, Massachusetts 02451
USA
Tel : 800-762-4000, Fax : 001 203-944-4904
Email: CustomerCareUS@perkinelmer.com

AMPEROMETRIC CELLS

System Specifications of Model 5040 Amperometric Analytical Cell, Dionex, USA

Operating in either traditional amperometric or pulsed modes with a choice of reference and working electrodes, our cells are ideal for typically chromatography methods requiring high-sensitivity detection of various redox-active analytes in applications ranging from neurochemical to carbohydrate analysis.

Ref: http://www.esainc.com/products/type/hplc_systems/cells/amperometric

Vendor (Amperometric cells)

Dionex
1228 Titan Way
P.O. Box 3603, Sunnyvale, CA
94088-3603, United States
Phone: 14087370700; Fax: +1408730 9403
Email: OrderAdministration@dionex.com

AUTOMATED ENZYME IMMUNOASSAY ANALYZER

An automated analyser is a medical laboratory instrument designed to measure different chemicals and other characteristics in a number of biological samples quickly, with minimal human assistance. Antibodies are used by some analysers to detect many substances by immunoassay and other reactions that employ the use of antibody-antigen reactions. When concentration of these compounds is too low to cause a measurable increase in turbidity when bound to antibody, more specialized methods must be used.

Ref: http://en.wikipedia.org/wiki/Automated_analyser#Immuno-based_analysers

System Specifications of AIA®-600II Automated enzyme immunoassay analyzer, Tosoh, Japan

Capacity   60 tests/hour
Interface   Bidirectional RS232C
Detector system   Top-to-top fluorescence detection
Assay methodology   One-step and two-step sandwich and competitive FEIA

Ref: http://www.tosoh.com/Products/tcdsci1.htm#ProductnameAIA6002

Vendor (Automated Enzyme Immunoassay Analyzer)

Tosoh Corporation
Shiba-koen First Bldg.,
3-2-8 Shiba, Minato-ku,
Tokyo 105-8623,
Japan
Tel: 81-3-5427-5179; Fax: 81-3-5427-5219
E-mail:info@tosoh.com

AUTOMATED PIPETTING SYSTEM

System Specifications of epMotion® 5075 TMX, Eppendorf AG, Germany

The epMotion 5075 TMX is the most flexible member of the epMotion family of automated pipetting systems. It includes an integrated TMX module to shake, heat or cool samples. The TMX module can be loaded and unloaded with the gripper. Pipetting to the mixer is possible before and after mixing steps. The TMX module is fully software-controlled enabling pipetting on other worktable positions while the mixer is in operation.

Labspace System	Width	140 cm / 56"
	Depth	75 cm / 30"
Dimensions		
Device	Width	107 cm / 43"
	Depth	61 cm / 25"
	Height	67 cm / 27"
Control panel	Width	25 cm / 10"
	Depth	15 cm / 6"
	Height	11 cm / 4.4"
Weight Device	85 kg / 188 lb	

Control panel	1.2 kg / 3 lb
Ambient conditions	
Air temperature	storage
operation	+15 to +35 °C / +59 to +95 °F
Atmospheric humidity	storage 10 to 80 % rF / rH
operation	55 to 75 % rF / rH
Atmospheric pressure	storage 300 to 1,060 hPa / 4.4 to 15.4 psi
operation	970 to 1,060 hPa / 14.1 to 15.4 psi
Mixing frequency	0, 300 – 2000 rpm, optimized for labware and application
Mixing amplitude	3 mm
Temperaturerange	–15°C below RT to 95°C measured in labware at 75% filling height
Temperaturehomogeneity	≤10% on the block
Heating rate	5°C/min, on the block
Cooling rate	3°C/min, on the block, above RT

Ref: http://www.eppendorf.com/int/index.php?pb=4db9308847b11b5b&action=products&contentid=1&catalognode=69733&productpage=3

Vendor (Automated Pipetting System)

Eppendorf AG
Barkhausenweg 1, 22339 Hamburg
Germany
Phone: +4940538010
Email: info@eppendorf.com

AUTOMATIC AMINO ACID ANALYSER AAA 400

System Specifications of Automatic amino acid analyser AAA 400, INGOS s.r.o., Czech Republic

AAA 400 Automatic Amino Acid Analyser is used for determination of amino acids in protein and peptide hydrolysates, for the dermination of free amino acids in physiological liquids and extracts as well. These characteristics determine AAA 400 for wide use in basic biochemical research of proteins, in research of human and animal nourishment, in medical diagnostics, in testing of drugs etc. Here is at disposal collection of examples of analyses in format pdf. To view the example analyses is necesary Adobe Acrobat Reader, which can be d ownloaded free of charge from Adobe company sites. The analysis is based on medium pressure ion-exchange chromatography with the colorimetric detection.

Sensitivity:	<50 pmol (S/N=5)
Reproducibility:	Retention times: 0.3 %(arg)
Peak area:	1% (ser, gly, his)
Range of use:	up to 150 ninhydrine positive substances physiological standard complete separation of 45 amino acids
Glass column	Ø3.7 x 450 mm
- standard life	~8.000 - 10.000 samples
Stainless steel column	to order
- standard life	~400 - 1.000 samples
Column temperature control:	20°C – 95°C (increment 0.1°C) (fast heating and cooling)
Exchange resins:	
hydrolyzates	Ostion LG ANB
free amino acids	Ostion LG FA
Pump system:	high pressure non-pulsation pumps
(ninhydrine and buffer pump)	(dual plunge, dual flying pistons)
flow range:	0.01 - 20 ml/min (blocked by SW to max. 10ml/min)
pressure range:	0 - 40 Mpa (blocked by SW to max. 20MPa)
Autosampler with sample cooling:	
option A	25 x 1.5 ml
option B and C	40 and 80 x 0.5 ml
Two channel photometer:	440 and 570 nm
flow cell volume:	standard 5 µl
Ninhydrine reactor:	ambient - 150 °C
Cooled ninhydrine storage:	2 l
Pre-column:	longlife ammonia filtr (Ostion KS 0804)
Nonstop running without attendance:	~7 days
Controling and data processing:	CHROMuLAN free license software (Win NT/2000/XP) download
Comunication:	RS 485
Power supply:	230 V ±10%, 50 Hz

Power input:
-starting   380 VA
-operating   230 VA
-standby   150 VA
Dimensions (w x h x d):   700 x 600 x 550 mm
Weight: 68 kg

Ref: http://www.instruments.ingos.cz/en/instrument-detail.php?id=automatic-aminoacid-analyser-aaa-400

Vendor (Automatic Amino Acid Analyser)

INGOS s.r.o.
Laboratory Instruments Division
K Nouzovu 2090
143 16 Praha 4
Czech Republic
Tel: +420 296 781 692, Fax: +420 244 403 051
E-mail:instruments@ingos.cz

CENTRIFUGE/VORTEX MULTISPIN

System Specifications of MSC3000, Centrifuge/Vortex Multispin, Biosan, Latvia

Centrifuge/vortex MultiSpin is a product of many year evolution of spinmixspin technology that is intended for collecting micro volumes of reagents on the microtube bottom (first centrifugation spin), following mixing (mix) and collecting the reagents again from the walls and cover of the microtube (second spin). Multi-Spin is a fully automatic device for reproducing sms -algorithm for 12 tubes at one time, that allows saving time considerably.

Multi Spin is four devices combined in one:

1. Centrifuge   Maximum RCF up to 800xg
2. Vortex   (3 mixing modes— soft, medium, hard; regulated time;
                       Vortexing regulation timer 1- 20 sec)
3. Centrifuge/vortex;
4. SMS-cycler for realization of the "sms algorithm"

Speed regulation range (increment 100 rpm)   1000–3500 rpm
RCF max.   800 x g
Spin timer   1 sec–99 min
Mixing strength   Soft, medium, hard
Mixing time   0–20 sec (increment 1 sec)
SMS-cycle regulation   1 - 999 cycles
Display   LCD, 2 x 16 signs
Safety   Autostop at open lid
Overall dimensions   190x235x125 mm
Weight, not more   2.5 kg
External power supply   AC 12V, 1.25 A

Ref: http://www.biosan.lv/eng/index.php?page=msc-3000

Vendor (Centrifuge/Vortex Multispin)

BioSan Ltd
7 Ratsupites Str., 2 build.
LV-1067, Riga, Latvia
Tel: 371-6-7426137; Fax: 371-6-7428101
E-mail: marketing@biosan.lv

CHLOROPHYLL FLUOROMETER

Chlorophyll fluorescence is light that has been remitted after being absorbed by chlorophyll molecules of plant leaves. By measuring the intensity and nature of this fluorescence, plant ecophysiology can be investigated. The development of portable fluorometers has allowed chlorophyll fluorescence to become a common method of measuring plant stress in plant ecophysiology studies. Chlorophyll fluorescence has been revolutionized by the use of modulated chlorophyll fluorometers in which the light source is modulated (rapidly switched on and off) and the detector is tuned to detect only fluorescence excited by the measuring light. This means the relative yield of fluorescence can be measured in the presence of background light. Crucially, this means chlorophyll fluorescence can be measured in the field in full sunlight.

Ref: http://en.wikipedia.org/wiki/Chlorophyll_fluorescence#Chlorophyll_fluorometers

System Specifications of OS-30p chlorophyll fluorometers by Opti-Sciences, Inc, USA

Test Modes	Fv/Fm
	Fast Kinetics - Basic OJIP
Excitation Source	Solid state 660nm source
	Modulation beam adjustable .2 to 1.0 uE
	Saturation/Actinic Intensity adjustable from
	100 ~ 3,000 uE depending on test mode
Detection method	Pulse Modulated w/ High Resolution
	sampling mode for Kautsky Induction curve recording.
Detector and Filters	A PIN photodiode with a 700 ~ 750 nm bandpass filter.
Sampling Rate	Variable from 10 to 30,000 points per second depending on phase of test.
Test Duration	Adjustable from 2 seconds ~ 255 seconds
Storage Capacity	512 Kb battery backed up RAM, supporting up to 8,190 test data
	sets, and 32 traces of up to 255 seconds each.
Digital Output	RS-232 port.(USB adapters are available)
User Interface	*Display:* 128 x 64 dot backlit graphics display
	Keyboard: 13 keys dedicated function layout
	Fiber Optic Cable Option: 1 meter in length
Power Supply	Rechargeable NiMH Battery Pack with charger
Battery Life	8 hours
Dimensions	18 cm x 7 cm x 6 cm

Ref: http://www.optisci.com/datasheet/os-30p.pdf

Vendor (Chlorophyll Fluorescence)

Opti-Sciences Inc.
8 Winn Avenue Hudson
NH 03051 USA
Ph: (603)883-4400; Fax: 603-883-4410
E-mail: sales@optisci.com

CLINICAL CHEMISTRY ANALYZER

Chemistry analyzer is used for determining thermostabile proteins.

System Specifications of B-200 Clinical Chemistry Analyzer, Mindray, China

• 200 tests per hour, up to 330 tests per hour with ISE (K, Na, Cl)
• 24-hour refrigeration for reagent tray
• Built-in bar code scanner
• Independent mixing stirrer
• Robust and user-friendly operation software
• Multi-language version available
• Pre-dilution and post-dilution for sample
• Bi-directional LIS interface

Ref: http://www.mindray.com/en/products/37.html#feature

Vendor (Clinical Chemistry Analyzer)

Mindray
Mindray Building, Keji 12th Road South,
High-tech Industrial Park,
Nanshan, Shenzhen 518057, P. R. China.
Ph: 86 (755) 26582888; Fax: +86 (755) 26582500
Email: intl-market@mindray.com

COULOMETRIC CELLS

System Specifications of Model 6210 Coulometric 4-channel Array Cell, Dionex, USA

ESA's proprietary high-efficiency cells are the ideal choice fo r the most- demanding applications, particularly where a high degree of selectivity, stability, and sensitivity is required. We design our coulometric cells with an array of uses in mind – from high-sensitivity detection and multiple-analyte analysis to simple sample conditioning. The CoulArray multi-electrode cell provides 4 independent working electrodes in one small package. This schematic shows a "cutaway" of a 4 channel ESA CoulArray cell. The CoulArray detector is available with up to 16 independent controlled electrodes.

Features and benefits

- Unparalleled selectivity – coulometric technology enables detection and analysis of compounds of interest in the presence of extraneous material that would interfere with detection by other electrochemical techniques.
- Unmatched performance – flow-through electrode design provides complete high-efficiency oxidation for maximum sensitivity.
- Minimal maintenance – robust graphite electrodes require little in the way of routine cleaning and regeneration procedures; solid-state reference electrode rarely needs maintenance.

Ref: http://www.esainc.com/products/type/hplc_systems/components/pumps/584

Vendor (Coulometric Cells)

Dionex
1228 Titan Way
P.O. Box 3603, Sunnyvale, CA
94088-3603, United States
Phone: 14087370700; Fax: +1408730 9403
Email: OrderAdministration@dionex.com

DOUBLE PULSE LASER INDUCED BREAKDOWN SPECTROSCOPY (DP-LIBS)

The experimental setup for dual-pulse LIBS (Fig. 1) was similar to that reported by Weidman et al (2009). It utilized two lasers: a Q-switched Nd:YAG laser (Brilliant —Quantel) operated at the fundamental wavelength of 1064 nm and capable of producing over 300 mJ of pulse energy in a 5 ns FWHM pulse duration and a TEA CO2 laser (Lumonics) operating at a wavelength of 10.6 μm, with 100 mJ of pulse energy in a 100 ns pulse followed by a 1 μs long tail.

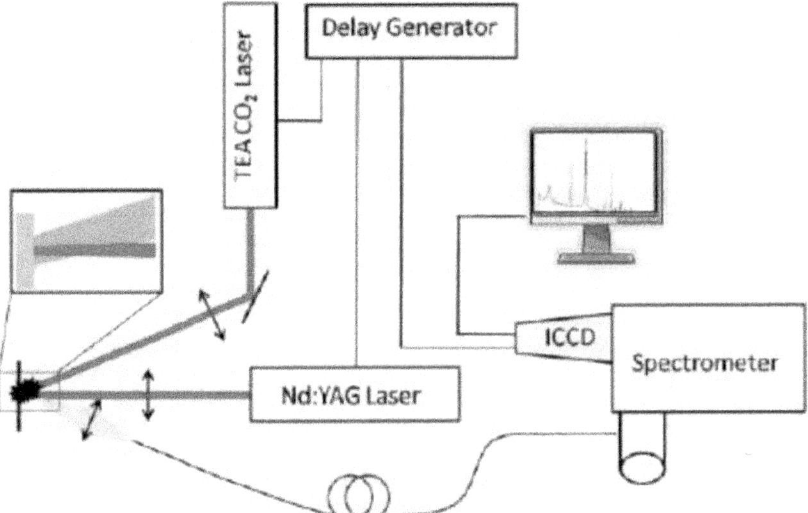

Figure: The double-pulse experimental setup

The pulse energy of the Nd:YAG laser could be varied using a ha lf-wave plate before a polarizing beam-splitter and was set to 17.5 mJ per pulse as measured using a pyroelectric energy meter (QE-25/Solo2 —Gentec). A 10 cm focal length lens was used to focus the Nd:YAG beam to a waist with a diameter of 400 μm, as determined using the knife-edge scan technique, giving an irradiance of 2.8 GW·cm- 2. The pulse energy from the CO2 laser could be varied using a set of polarizers and the energy was adjusted to 63 mJ per pulse, measured using the same pyroelectric energy meter. A 25 cm focal length germanium lens wa s used to focus the CO2 laser on the sample that was 21 cm from the lens. A defocused spot diameter of 1.2 mm, also determined using a knife edge scan, provided an irradiance of 5 MW·cm- 2 on the target. A larger CO2 spot allowed greater energy to be delivered for a given target irradiance. In the case examined here, the CO2 laser alone did not create a plasma. The CO2 laser beam was at an angle of approximately 15° with respect to the surface normal and the Nd:YAG was normally incident. Timing synchronization between both lasers was accomplished using two pulse-delay generators: a DG-645 (Stanford Research Systems) set as the master clock and a DG-535 (Stanford Research Systems) to control the delay between lasers and detector gate. The experiment was run at 0.3 Hz repetition rate, as limited by the CO2 laser. The interpulse delay was measured between the rising edge half-maxima of the two pulses. The detector gate delay was measured with respect to the rising edge half-maximum of the second pulse.

Weidman M (2009) Thermodynamic and spectroscopic properties of Nd:YAG-CO2 Double-Pulse Laser-Induced Iron Plasma, Spectrochimica Acta Part B: Atomic Spectroscopy

Weidman M, Matthieu Baudelet, Santiago Palanco et al. (2010) Nd:YAG-CO2 double-pulse laser induced breakdown spectroscopy of organic films. Optics Express 18(1):259-266

Source: http://www.dtic.mil/cgi-bin/GetTRDoc?Location=U2&doc=GetTRDoc.pdf&AD=ADA519506

Q-SWITCHED ND :YAG COMPACT OSCILLATORS (QUANTEL, FRANCE)

The latest technological advances have given it even greater stability, complete remote control, active temperature stabilization and, last but certainly not least, a 5th harmonic generator. Brilliant is also a complete line of related products: the Brilliant Eazy and Brilliant B (compact oscillators), their harmonic generators, the Twins (double-pulse systems) and the Rainbow (solid-state tunable laser).

Options and accessories

- 2nd, 3rd, 4th and 5th harmonic generation
- High Energy UV option
- Intra-cavity etalon
- Eye-safe OPO
- Multimode resonator
- Beam attenuator
- Power meter
- PIV configuration

System Specifications

Type de résonateur	Gaussien (GRM)	Brilliant		
Repetition Rate (Hz)		10	20	50
Energy (mJ)	1064 nm	360	350	150
	532 nm	180	160	65
* high energy option	355 nm	65 / 100*	60 / 70*	20
	266 nm	40	30	12
	213 nm	8	6	2
Pulse Duration (ns) @ 1064nm		≈5	≈5	≈6
Energy Stability (%)	Variation/Emoy 100% shots	+/- 2%	+/- 2%	+/- 3%
Divergence (mrad) On 200 shots, full angle at $1/e^2$		0.5	0.5	0.7
Beam Diameter (mm)		≈6	≈6	≈6
Dimensions	Laser Head	136 × 80 × 469		
H × w × L in mm	Integrated Cooling & Electronics	585 × 592 × 286		

Ref: http://www.quantel.fr/industrial-scientific-lasers/uk/produit-53-brilliant.html

Vendor (DP-LIBS)

Quantel
BP23, 2bis Avenue du Pacifique
91941 Les Ulis Cedex
France
Ph : +33(0)169291716; Fax : +33(0)169291706
E-mail: quantel@quantel.fr

DUAL METER PH / CONDUCTIVITY (S47 SevenMulti™)

SevenMulti™ basic device with two expansion units for pH and conductivity measurement. Including power supply, electrode arm and base, operating instruction, test certificats and declaration of conformity.

Specifications - S47 SevenMulti™ dual meter pH / conductivity (Mettler-Toledo Inc., USA)

pH-range	-2.000 to 19.999
pH-resolution	variable: 0.001 / 0.01 / 0.1
pH-relative accuracy	± 0.002
mV-range	-1999 to 1999
mV-resolution	0.1
mV-relative accuracy	± 0.1
Temperature range °C	-30.0 to 130.0
Temperature resolution °C	0.1
Temperature accuracy °C	± 0.1
Display	Custom LCD
Cond. accuracy	+/- 0.5%
Cond. resolution	0.001 µS/cm to 1000 mS/cm, autoscaling
TDS range	0.1 mg/L ... 1000 g/L
Salinity range	0.00 ... 80.00
Resistivity range	0.00 ... 20 MOhm*cm

Ref: http://us.mt.com/us/en/home/products/Laboratory_Analytics_Browse.html

Vendor (Dual Meter pH / Conductivity)

Mettler-Toledo Inc.,
1900 Polaris Parkway, Columbus
OH 43240, USA
Ph: + 800 638 8537
Email: Not Available

ELECTROCHEMICAL ANALYZER/WORKSTATION (MODEL 600D, CH INSTRUMENTS, INC.,)

The Model 600D series is designed for general purpose electrochemical measurements. The figure below shows the block diagram of the instrument. The system contains a fast digital function generator, high speed data acquisition circuitry, potentiostat, and a galvanostat (available only in select models). The potential control range is ±10 V and the current range is ±250 mA. The instrument is capable of measuring current down to tens of picoamperes. The steady state current of a 10 µm disk electrode can be readily measured without external

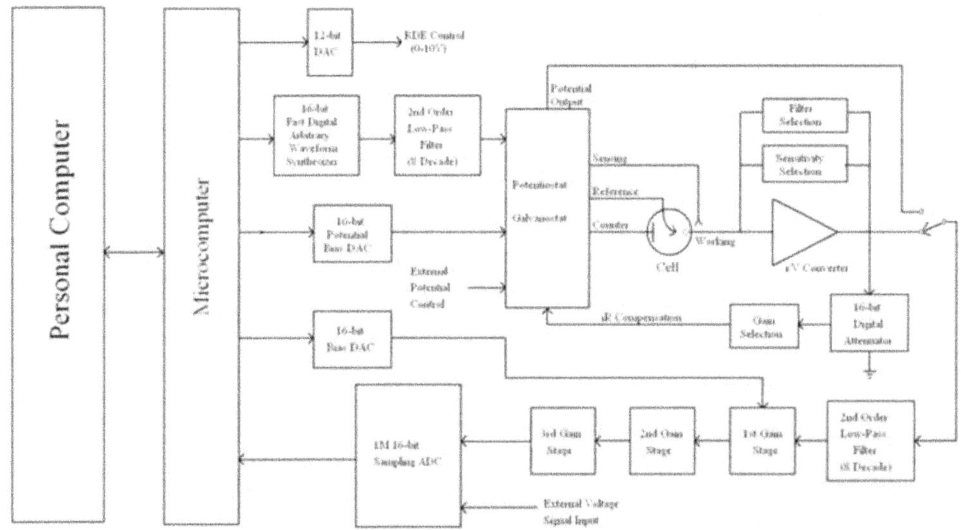

adapters. With the CHI200B Picoamp Booster and Faraday Cage (fully automatic and compatible with the CHI600D series), currents down to 1 pA can be measured. These instruments are very fast. The function generator can update at a 10 MHz rate, and the maximum sampling rate is 1 MHz at 16-bit resolution. The instrument offers a very wide dynamic range of experimental time scales. For instance, the scan rate in cyclic voltammetry can be up to 1000 V/s with a 0.1 mV potential increment or 5000 V/s with a 1 mV potential increment. The potentiostat/galvanostat uses a 4-electrode configuration, allowing it to be used for liquid/liquid interface measurements and eliminating the effect of the contact resistance of connectors and relays for high current measurements. The data acquisition systems allow an external input signal (such as spectroscopy signals to be recorded simultaneously with electrochemical data). The instrument will also automatically re-zero both potential and current, so that periodic re-calibration of the instrument can be avoided.

The 600D series is the upgrade to our very popular 600/600A/600B/600C series. The instrument utilizes flash memory, allowing instrument updates to be distributed electronically instead of the inconvenient shipment of an EPROM chip. The 600D series has a USB port (default) and a serial port for data communication with the PC. You can select either USB or serial (but not both) by changing the switch setting on the rear panel. The 600D series also includes a true integrator for chronocoulometry. A 16-bit highly stable bias circuitry is used for current or potential bias. This allows wider dynamic range in ac measurements. It can also be used to re-zero the dc current output. The model 600D series can be upgraded to a bipotentiostat. The model 700D series will be an add-on board to the 600D series. It will therefore be identical to the 600D series when used for single channel measurements. When it is used as a bipotentiostat, the second channel can be controlled at an independent constant potential, to scan or step at the same potential as the first channel, or to scan with a constant potential difference with the first channel. The second channel is available for many voltammetric and amperometric techniques. The instrument also provides various electrochemical techniques, integrated digital CV simulator, and an impedance simulation and fitting program. These features provide powerful tools for both electrochemical mechanistic studies and trace analysis. We provide several different models in the 600D series. The model comparison table compares the different models. Other than what is listed, the specifications and features of these models are identical. Models 600D and 610D are basic units for mechanistic study and electrochemical analysis, respectively. They are also great for teaching purposes. Models 602D and 604D are for corrosion studies. Models 620D and 630D are comprehensive electrochemical analyzers. Models 650D and 660D are advanced electrochemical workstations.

Ref: http://www.chinstruments.com/chi600.shtml

System Specifications of Model 600D Series Electrochemical Analyzer/Workstation

Potentiostat
Galvanostat (Model 660D)
Potential range: -10 to 10 V
Potentiostat rise time: < 1 µs
Compliance voltage: ±12 V
3- or 4-electrode configuration
Current range: 250 mA
Reference electrode input impedance: 1x1012 ohm
Sensitivity scale: 1x10-12 - 0.1 A/V in 12 ranges
Input bias current: < 50 pA
Current measurement resolution: < 0.01 pA
Minimum potential increment in CV: 100 µV
Potential update rate: 10 MHz
Fast data acquisition: 16 bit @ 1 MHz
External signal recording channel
Automatic and manual iR compensation
Automatic potential and current zeroing
Flash memory for quick software update
Serial port or USB selectable for data communication

CV and LSV scan rate: 0.000001 to 5000 V/s
Potential increment during scan: 0.1 mV @ 1000 V/s
CA and CC pulse width: 0.0001 to 1000 sec
CA and CC Steps: 320
True integrator for CC
DPV and NPV pulse width: 0.001 to 10 sec
SWV frequency: 1 to 100 kHz
ACV frequency: 0.1 to 10 kHz
SHACV frequency: 0.1 to 5 kHz
IMP frequency: 0.00001 to 100 kHz (extend to 1MHz for impedence 10-1000 ohm)
IMP amplitude: 0.00001V to 0.7V RMS
Potential, current low-pass filters, covering 8-decade frequency range, Automatic and manual setting
RDE rotation control voltage output: 0 - 10 V (Model 630D and up)
Cell control: purge, stir, knock
Maximum data length: 128K-8192K selectable
Dimensions: 14.25"(W) x 9.25"(D) x 4.75"(H)
Weight: 12 Lb.

Vendor (Electrochemical Analyzer/Workstation)

CH Instruments, Inc.
3700 Tennison Hill Drive
Austin, TX 78738-5012
USA
Tel: (512) 402-0176; Fax: (512) 402-0186
E-mail: info@chinstruments.com

ELECTROPHORETIC ANALYZER

System Specifications of Electrophoretic Analyzer EA102, Villa Labeco spol. s. r.o., Slovakia

The analyzer is possible to use both contact and contactless conductivity column detectors with this analyser and the whole unit is controlled by up-graded, easy to operate Windows software. ELECTROPHORETIC ANALYSER EA 102 in the 2-dimensional column arrangement for ITP-ITP (isotachophoresis- isotachophoresis) or ITP-CZE (isotachophoresis- capillary zone electrophoresis) analysis. Configuration with preseparation column (using a contactless conductivity detector), micropreparative valve (to enable concentration and collection of very pure sample fractions), and analytical column with contact conductivity detector.

Dimensions: 275 x 620 x 256 mm (w x h x d)

Ref: http://www.villalabeco.sk/eng_ponuka_EA102.htm

Vendor (Electrophoretic Analyzer)

Villa Labeco spol. s. r.o.
Chrapčiakova 1. 052 01 Spišská Nová Ves
Slovakia
Tel: 00421 53 44 260 32; Fax: 00421 53 4196
104
Email: villa@spisnet.sk

FLUORESCENCE IMAGING SYSTEM

A fluorescence microscope is an optical microscope used to study properties of organic or inorganic substances using the phenomena of fluorescence and phosphorescence instead of, or in addition to, reflection and absorption.

System Specifications of DynaMyc - Lifetime Microscope System, Horiba Scientific, Japan

Microscope	Based on upright Olympus BX51 microscope	**Detector**	Picosecond photon detection module
Objectives	Plan achromat x10 and x50, other magnifications available.	Spectral range	185-650 nm / 300-850 nm
Confocal pinhole	5 position turret.	Dark count	< 80cps
	4 diameters, from 100 µm to 1000 µm, Motorized, computer controlled		
Camera		**Filters**	
Color USB CCD	2 Mpix, 8 bits	Excitation filters	10 nm BP filters
Fluorescence camera	1.4 Mpix cooled, optional	Fluorescence filters	2 positions
			Motorized, computer-controlled
		Emission filters	2 positions: 40 nm BP filters
			Motorized, computer-controlled
		ND filters	6 positions: 0, 0.3, 0.6, 1, 2 and 3 OD
			Motorized, computer-controlled
Excitation sources	Fiber-coupled pulsed laser sources	**Software**	
Repetition rate		Data acquisition	DataStation software
Wavelength range	10 kHz to 20 MHz, with PicoBrite sources	Data analysis	DAS6 software, inc. optional
	From 375 nm to 670 nm.	Operating system	reconvolution
			Windows XP / Windows Vista
Motorized stage		**Dimensions**	140 cm x 90 cm x 80 cm
Resolution	0.5 µm		
Travel range	75 x 50 mm		
Manual control	With joystick		
Automatic control	Through DataStation software		
TCSPC electronics	Single-photon counting detection		
Lifetime range	100 ps to 10 µs, depending on sample		

Vendor (Fluorescence Imaging System)

HORIBA Ltd
Mr Makoto Hashimoto
2 Miyanohigashi, Kisshoin
Minami-Ku 601-8510 KYOTO Japan
Ph: +81 75 313 8121, Fax: +81 75 321
5725
Email: makoto.hashimoto@horiba.com

HANDHELD POTENTIOSTAT/GALVANOSTAT

System Specifications of EmStat potentiostat, PalmSens, The Netherlands

EmStat is the smallest USB potentiostat available. It is available in two versions, EmStat and EmStat2, and has six or eight current ranges from 1 nA to 100 uA or 10 mA full scale re spectively, with a minimum resolution of 1 pA. The potentiostat is controlled by means of either a

USB cable or optionally by means of a serial (RS-232) cable. The USB version of EmStat is powered by the USB port. It provides the most relevant electroanalytical measurement techniques and is used with the PSTrace program.

- dc-potential range	± 2 V
- compliance voltage	± 5 V
- dc-potential resolution	EmStat: 1 mV
	EmStat2: 0.1 mV
- potential accuracy	$\leqslant$0.2 % offset and 0.2 %
- current ranges	EmStat: 1 nA to 100 µA (6 ranges)
	EmStat2: 1 nA to 10 mA (10 ranges)
- maximum current	$\pm$ 10 mA
- current resolution	0.1 % of current range
	1 pA on lowest current range
- accuracy	$\leqslant 0.2$ % of current range at 100 nA to 100 µA
- electrometer amplifier input	> 100 Gohm // 4 pF
- rise time < 50	µs
Dimensions	62 mm x 40 mm x 25 mm
Power	By USB port or optionally 5 V adapter
Interfacing	USB (optionally RS-232 or serial TxD/RxD)
Minimum requirements	Windows XP or Vista

Ref: http://www.palmsens.com/instruments/emstat/specifications/

Vendor (Handheld Potentiostat/Galvanostat)

Palm Instruments BV
Ruitercamp 119
3992 BZ Houten
The Netherlands
Tel.: +31302459211; Fax.: 31302459212
Email: info@palmsens.com

HANGING MERCURY DROP ELECTRODES (HMDE)

The hanging mercury drop electrode (HMDE) is a working electrode variation on the dropping mercury electrode (DME). Experiments run with dropping mercury electrodes are referred to as forms of polarography. If the experiments are performed at an electrode with a constant surface (like the HMDE) it is referred as Voltammetry. Mercury fl ow through the capillary is controlled by means of a closing needle actuated by an electromagnetic relay. The timing circuit is very simple and consists only of resistors and capacitors. This simple electrode could find various applications in the teaching laboratory for illustrating the principles of general electrochemistry, stripping voltammetry, and complex equilibria in solutions.

Ref: http://pubs.acs.org/doi/abs/10.1021/ed077p98

System Specifications of BASi Controlled Growth Mercury Electrode (BASi, USA)

Cell Modes	CGME, SMDE, and DME
Operations	Manual and remote control of mercury dispense; drop knocker; stir on/off; gas purge/blanket
Magnetic Stirrer	16 selectable rates from 50-800 rpm
Mercury Reservoir	Sealed reservoir minimizes opportunity for mercury spills. Minimal use of metal parts within reservoir or capillary to avoid metal contamination.
Mercury Degassing	Mercury degassing kit is included with purchase and includes a hand-actuated vacuum pump and accessories
Spill Management	Spill tray and mercury pickup tool are included in the standard package
Standard Capillary	150 µm (0.006 ") ID. Large diameter reduces plugging and loss of electrode contact. Beveled tip design
Gas Purging	5 psi max. Needle valve controlled.
Cell Top	Separate port in cell top for dispensing sample or reagents. Accommodates BASi voltammetry electrodes for use as a cell stand for voltammetry experiments.
Options	100 µm ID capillary
	Faraday cage
	Small volume and water-jacketed cell vials
Physical	42 x 18 x 23 cm (H x W x D), 6 kg

Mercury drop electrodes are generated by the BASi Controlled Growth Mercury Electrode (CGME) in three modes:

- DME (Dropping Mercury Electrode) - mercury is allowed to flow freely from the reservoir down the capillary and so the growth of the mercury drop and its lifetime is controlled by gravity. (The optional 100 µm capillary is recommended for this mode.)

- SMDE (Static Mercury Drop Electrode) - the drop size is determined by the length of time for which the fast-response capillary valve is opened, and the drop is dislodged by a drop knocker. The dispense/knock timing is microprocessor-controlled and is typically coordinated with the potential pulse or square-wave waveform. This mode can also used to generate the Hanging Mercury Drop Electrode required for stripping experiments.

- CGME (Controlled Growth Mercury Electrode) - the mercury drop is grown by a series of pulses that open the capillary valve. The number of pulses, their duration, and their frequency can be varied by PC control, providing great flexibility in both the drop size and its rate of growth. This CGME mode can be used for both polarographic and stripping experiments.

Ref: http://www.basinc.com/products/ec/cgme.php

Vendor (HMDE)

BASi Corporate Center
Purdue Research Park
2701 Kent Avenue
West Lafayette, IN 47906
USA
Ph: 800 845 4246, Fax: 765 497 1102
Email: http://www.basinc.com/ask/

HPLC AUTOSAMPLER

System Specifications of Model 542 HPLC Autosampler, Dionex, USA

Feature-rich and incorporating an integrated column oven, the Model 542 is a cost-effective autosampler for HPLC analysis. This compact-sized unit offers many features found in other more expensive autosamplers. It incorporates three selectable injection modes: full loop, partial loopfill, and microliter pick-up. Sample processing capabilities are provided for sample mixing, dilution, and on-line derivatization.

Sample Capacity:	Standard tray: 84 vials of 1.8 mL (Std) and 3 vials of 10 mL (LSV)
Vial Dimensions (including cap):	Maximum vial height: 47 mm, Minimum vial height: 32 mm
Loop Volume:	I - 5000 µL (interchangeable)
Dispenser Syringe:	250 µL (Std) or 1000 µL syringe
Vial Detection:	Missing vial detected by vial sensor
Headspace Pressure:	Built-in compressor
Switching Time Injection Valve:	Electrically <100 msec
Piercing Precision Needle:	± 0.6 mm
Wash Solvent:	External wash solvent bottle
Bio-compatible Valve:	PEEK' injection valve, with quick connect mounting
Wetted Parts in Flow Path:	SS316, PTFE, TEFZEL, VESPEL, Glass, Teflon, PEEK
Built-in Column Oven:	Temperature Range: ambient +5 to 60 C ± II C
Oven Dimensions:	I" W x I" D x 11.811 L (2.5 cm x 2.5 cm x 30 cm)
Communication Port:	Serial RS232C
Output:	Injection marker, 2 Auxiliary outputs, Alarm output (Relay & TTL)
Input:	Next injection, freeze & stop (TR)
Ambient Environment:	10-40'C, 5-95% relative humidity, non-condensing
Unit Dimensions:	30 cm x 50 cm x 34 cm (11.8" H x 19.7" W x 13.4" D)
Weight:	19 kg (42 lb.), 21 kg (46 lb.) with cooling option
Power Requirements:	115/230 VAC; ± 15%; 50/60 Hz; 200 VAC
Product Certification:	CE
PERFORMANCE	
Reproducibility:	RSD < 0.3% for full loop injections RSD < 0.5% for partial loop fill injections, injection volumes > 10 µL RSD < 1.0% for µL pick-up injections, injection volumes > 10 µL

Carry-Over:	< 0.05% with programmable needle wash

PROGRAMMING

Methods:	9 programmable methods with priority sample mode
Injection Modes:	Full loop injection, Partial loop fill injection, µL pick-up injection
Injection Volume:	I µL - 5000 µL, with I µL increments
Injections per Vial:	Maximum 9 injections
Analysis Time:	Maximum 9 hr. 59 min. 59 sec.
Programmable Needle Wash:	Wash between inj ections, between vials, between series
Programmable Timed Events:	4 x ALJX-1 on/off, 4 x AL7X-2 on/off, Initial oven setpoint, 2 x new oven setpoint
Series:	9 programmable series

SAMPLE TRAY COOLING

Temperature Control:	Integrated, Peltier type
Temperature Range/Capacity:	4-15? C, no greater than 20?' C below ambient
Temperature Accuracy:	± 2? C

Ref: http://www.esainc.com/products/type/hplc_systems/components/autosamplers/542

Vendor HPLC Autosampler

Dionex
1228 Titan Way
P.O. Box 3603, Sunnyvale, CA
94088-3603, United States
Phone: 14087370700; Fax: +1408730 9403
Email: OrderAdministration@dionex.com

HPLC COLUMN

System Specifications of Kinetex ® core-shell, Phenomenex, USA

Kinetex ® core-shell 2.6 µm and 1.7 µm particles were engineered to provide increased efficiencies and improved performance compared to traditional fully porous particles.

- Increase resolution, throughput, and sensitivity
- Decrease solvent consumption
- Save time and money
- 1.7 µm and 2.6 µm particles are directly scalable

2.6 µm Kinetex columns on your HPLC or UHPLC systems provide sub-2 µm efficiencies with significantly lower back pressure.

Part Number:	00A-4462-AN
Description:	Kinetex™ 2.6 µm C18 100 Å LC Column 30 x 2.1 mm
Stationary Phase:	C18 with TMS endcapping
Solid Support:	Core-shell Silica
Format: Column	
Separation	Mode:Reversed Phase
Recommended Use:	Ultra-high performance on any LC system for hydrophobic compounds
USP Designation:	L1

Ref: http://www.phenomenex.com/Products/HPLCDetail/Kinetex/C18?returnURL=/Products/Search/HPLC

Vendor (HPLC Column)

Phenomenex
411 Madrid Avenue
Torrance, CA 90501-1430
USA
Tel: (310) 212-0555; Fax: (310) 328-7768
Email: info@phenomenex.com

System Specifications of ZIC®-pHILIC, Polymeric Column for Hydrophilic Compounds, Sequant-Merck, Sweden

The ZIC®-pHILIC Hydrophilic Interaction Liquid Chromatography column is the ideal choice for separation of polar and hydrophilic compounds as it is orthogonal to reversed phase chromatography (RPLC) columns. For example, compounds such as amino acids, peptides, carbohydrates, plant extracts and various other polar compounds that might have little or no retention in RPLC, generally have strong retention on the ZIC®-pHILIC column. The columns have a zwitterionic stationary phase on porous polymer particles and the separation is achieved by a hydrophilic interaction mechanism superimposed on weak electrostatic interactions. Due to the polymeric support particle, the column can be operated in a broad pH range, which can be used to improve the retention and selectivity for many compounds (100 x 4.6mm, 5 µm).

Ref: http://www.sequant.com/default.asp?ml=11523

Vendor (HPLC Column)

SeQuant
Tvistevägen 48
S-907 36 Umeå
Sweden
Ph: +46-90-154880;Fax: +46-90-154883
E-mail: info@sequant.com

HPLC SOLVENT DELIVERY SYSTEM

System Specifications of Model 584 Pump for CoulArray Clinical Systems, Dionex, USA

Optimized for HPLC electrochemical detection, the Model 584 Solvent Delivery Module offers pulse-free solvent delivery – even at micro-flow-rate ranges below 50 µl/min – and offers the following features and benefits:

Dual parallel plungers – provide low pulsation and greater seal life
Micro-volume plunger displacement – 10 µL per plunger stroke, improves limits of electrochemical detection
Flow rate range of 0.001 - 10 mL/minute – can be used in analytical, microbore, and LC-MS separations
Automatic seal wash – prolongs plunger seal life for greater reliability
Enhancement package – for reduced noise and lower limits of detection with all HPLC detectors
Electronic diagnostics – for easier maintenance
Improved fluidics – more reliable and easier to maintain
High-gradient accuracy and precision – for greater reproducibility

This LC pump is combined to create a dual high pressure gradient configuration using the CoulArray Gradient upgrade kit. (70-7057).

Pump:	Micro-volume double plunger (approximately 10 µL/stroke)
Modes of solvent delivery:	Constant flow or constant pressure solvent delivery
Flow-rate range:	0.0001 to 10.0000 mL/minute (10 to 340 bar)
Flow-rate accuracy:	No more than 1% or 0.5 µL/minute
Flow-rate precision:	0.3% maximum (RSD: 0.1% maximum)
Damper (external):	Internal capacity of approximately 0.8 mL
Materials in contact with mobile phase:	316 stainless steel, ruby, sapphire, Teflon®, PEEK, Hastelloy® C
Time program:	Commands for flow rate, pressure, event, loop (repeated programs), a total of 10 files and 320 steps
Pressure accuracy:	±10% or 1 mPa, whichever is greater
Communications:	External RS-232 converter, optional
Size:	10" (W) x 5" (H) x 16.5" (D) (260 mm x 140 mm x 420 mm), protruding components are not included
Weight:	22 lbs (10 kg)
Operating-temperature range:	41° to 95° F (5° to 35°C)
Power requirements:	90 to 240 VAC, 150 VA , 50/60 Hz
Certification:	USA, Canada, CE, FCC

Ref: http://www.esainc.com/products/type/hplc_systems/components/pumps/584

Vendor (HPLC Solvent Delivery System)

Dionex
1228 Titan Way
P.O. Box 3603, Sunnyvale, CA
94088-3603, United States
Phone: 14087370700; Fax: +1408730 9403
Email: OrderAdministration@dionex.com

HIGH-PERFORMANCE LIQUID CHROMATOGRAPHY (HPLC)

High-performance liquid chromatography (sometimes referred to as high-pressure liquid chromatography), HPLC, is a chromatographic technique that can separate a mixture of compounds and is used in biochemistry and analytical chemistry to identify, quantify and purify the individual components of the mixture.

Ref: http://en.wikipedia.org/wiki/High-performance_liquid_chromatography

System Specifications of The Agilent 1260 Infinity Series LC, USA

- A maximum pressure of 600 bar to utilize smaller particle size columns for higher resolution and faster separations.
- Configurable pump delay volume down to 120 µL together with a flow range up to 5 mL/min at 600 bar provides universal applicability from narrow bore (2.1 mm ID) to standard bore (4.6 mm ID) columns, matching the needs for both LC and LC/MS.
- New optical design in 1260 Infinity Diode Array Detector – including Agilent Max-Light cartridge flow cell – offers a new level of UV-sensitivity and baseline robustness, and fast spectral acquisition at data rates up to 80 Hz.
- Ultra sensitivity is achieved through revolutionary Agilent Max-Light cartridge cell with 60 mm optical path length
- (typically noise: $< \pm 0.6$ µAU/cm).
- Next generation flow-through design of the 1260 Infinity Autosampler achieves highest precision for both large and small injection volumes without changing sample loops.
- New 1290 Infinity LC Injector HTC/HTS is designed specifically to meet the throughput and reliability demands of critical LC/MS applications, offering lowest carryover with dynamic load and wash (DLW) principle.
- New pull-out valve drive and user-exchangeable valve heads in the 1290 Infinity Thermostatted Column Compartment boosts usability and paves the way for ultra high-throughput, multi-method and automated method development solutions.
- Broad range of more than 140 RRHT columns 1.8 µm for wide applicability and easy method transfer.
- New ZORBAX Rapid Resolution High Definition (1.8 µm) LC columns facilitate easy, fast and secure method transfer from HPLC to UHPLC through identical column chemistries.
- New Lab Advisor software – with intuitive diagnostic and monitor capabilities and alert functions to notify you of problems – helps you manage your lab for best chromatographic quality

Ref: http://www.chem.agilent.com/Library/brochures/5990-5061EN.pdf

Reversed phase chromatographic column Zorbax eclipse AAA C18

Agilent's ZORBAX Column Selection Guide for HPLC is the only chromatography guide that combines a broad offering of LC columns with expert advice that can help you achieve reliable results faster. AAA C18 is used for amino acid separation and analysis.

Ref: http://www.interlab.ru/articles/files/ZORBAX_Column_Selection_Guide2.pdf
http://www.chem.agilent.com/Library/datasheets/Public/5980-3088.pdf

Vendors (HPLC)

Agilent Technologies Inc.	**Waters Corporation**	**Shimadzu Scientific Instruments**
Santa Clara, California	34 Maple Street, Milford	7102 Riverwood Drive
Oswego, Illinois,	Massachusetts, 01757	Columbia, MD 21046
USA	USA	USA
Ph: 1-800-227-9770	Ph: 1 508 478-2000	Phone:1-800-477-1227, Fax: 410-381-1222
Email: agilent_inquiries@agilent.com	Email: emily_bouvier@waters.com	http://www.shimadzu.com/contact/index.html

HIGH RESOLUTION ULTRASONIC SPECTROSCOPY

Ultrasonic material analysis is based on the measurement of parameters of ultrasonic waves propagating through the sample. This provides information on the interaction of the ultrasonic waves with the sample's interior, thus enabling analysis of its physical and chemical properties. The ultrasonic wave is a wave of oscillating pressure and associated longitudinal deformation. The Ultrasonic Scientific HR-US spectrometers employ a principle where the path length of the ultrasonic wave in the sample exceeds the size of the sample. The use of modern advances in ultrasonic design, electronics and digital processing allow the attainment of ultrasonic measurements with record resolution (down to ±0.2mm/s for ultrasonic velocity) in a broad range of the sample volumes, down to a single droplet. Ultrasonic spectroscopy is simply spectroscopy employing sound waves. In particular, it uses a high frequency acoustical wave (similar or higher to those used by dolphins for communication and bats for navigation). The wave probes intermolecular forces in materials. Oscillating compression (and decompression) in the ultrasonic wave causes oscillation of molecular arrangements in the sample, which responds by intermolecular attraction or repulsion.

High Resolution Ultrasonic Spectroscopy specifications and list of vendors

System Specifications of High Resolution Ultrasonic Spectroscopy specifications (HR-US 102, Thermo Fisher Scientific products, USA)

The HR-US 102 spectrometer has evolved from the award winning HR-US 101 spectrometer. Compared to the HR-US 101, it has an extended frequency range, enhanced user control system and a digitally controlled stirring system. It is ideal for measuring a wide variety of liquid systems, from dilute solutions to concentrated emulsion. It can be used to monitor processes such as molecular structural changes, thermal transitions, chemical reactions, aggregation formation, crystallization etc. The digitally controlled stirring system also allows for titrations to be carried out to monitor binding reactions or concentration dependence.

Technical parameters of the model

- Two 1ml cells for parallel or differential sample analysis
- Digitally Controlled Stirring System
- Frequency Range from 2 to 18 MHz (cells with extended frequency range are available on request)
- High-Resolution & Reproducibility (down to ±0.2mm/s for Velocity and 0.2% Attenuation)
- Variety of Measurement Regimes:
  - o Temperature Ramp Mode
  - o Kinetic Mode
  - o Titration Mode
  - o Multi-Frequency Mode
- Temperature Range -20°C to 120°C (standard) and -40°C to 150°C (extended)
- Absolute Temperature Control down to 0.01°C

Ref: http://www.ultrasonic-scientific.com/Products/hrus_102.htm

Vendor (High Resolution Ultrasonic Spectroscopy)

Sonas Technologies Ltd
42 Ailesbury Grove,
Dundrum, Dublin 16
Ireland
Ph: + 353 1 2987567, Fax: +353 1 298
7567
E-mail: info@sonas-group.com

HYPERCHEM SOFTWARE

About HyperChem Professional Software

HyperChem is the newest Windows member of the HyperChem Family. Computational methods include molecular mechanics, molecular dynamics, and semi-empirical and ab-initio molecular orbital methods, as well as density functional theory. HyperChem Data and HyperNMR are included as part of HyperChem. New features are continually added and include elegant Open GL rendering, TNDO, RM1, Charmm protein simulations, molecules in magnetic fields, interfaces to third-party applications, calculations of structure, spectra, rate constants and much more. HyperChem is applicable to macromolecules as well as small molecules and is scriptable.

Ref: http://www.hyper.com/Prod ucts/tabid/354/Default.aspx

Vendor (Hyperchem Software)

HyperCube
1115 NW 4th St. Gainesville, FL 32601
USA
Ph: (352) 371 7744, (800) 960 1871
Email: sales@hyper.com

ICCD CAMERA

An intensified charge-coupled device (ICCD) is a CCD that is optically connected to an image intensifier that is mounted in front of the CCD. An image intensifier includes three functional elements: a photocathode, a micro-channel plate (MCP) and a phosphor screen. These three elements are mounted one close behind the other in the mentioned sequence. The photons which are coming from the light source fall onto the photocathode, thereby generating photoelectrons. The photoelectrons are accelerated towards the MCP by an electrical control voltage, applied between photocathode and MCP. The electrons are multiplied inside of the MCP and thereafter accelerated towards the phosphor screen. The phosphor screen finally converts the multiplied electrons back to photons which are guided to the CCD by a fiber optic or a lens.

An image intensifier inherently includes a shutter functionality: If the control voltage between the photocathode and the MCP is reversed, the emitted photoelectrons are not accelerated towards the MCP but return to the photocathode. Thus, no electrons are multiplied and emitted by the MCP, no electrons are going to the phosphor screen and no light is emitted from the image intensifier. In this case no light falls onto the CCD, which means that the shutter is closed. The process of reversing the control voltage at the photocathode is called gating and therefore ICCDs are also called gateable CCD cameras.

Ref: http://en.wikipedia.org/wiki/Charge-coupled_device#Intensified_charge-coupled_device

System Specifications of ICCD Camera PI-MAX3, Princeton Instruments, USA

PIMAX 3 is available in 1Kx1K (PI-MAX3: 1024i) and 1024x256 (PI-MAX3: 1024x256) formats, ideal for time-resolved imaging and spectroscopy applications. 1024i Camera achieves close to video frame rates (even at full 1k x 1k resolution). With ROI and binning, the cameras can achieve hundreds even thousands of spectra per second can be acquired while in spectroscopy mode; ability to capture a gated image or spectrum for every pulse of a high-repetition-rate laser. It is capable of 1MHz sustained repetition rate, which is 20x improvement over previous generation ICCD cameras. It quickly transfers a full-resolution image under the interline CCD pixel's adjacent mask area, enabling the camera to take a second frame in as little as 2μsec (limit imposed by P46 phosphor decay time). It has provision to cool the photocathode to reduce EBI by as much as 20 times. High-bandwidth (125MB/sec or 1000 Mbps) data interface for real-time image transmission; supports remote operation from more than 50m away.

Vendor (ICCD Camera)

Princeton Instruments
3660 Quakerbridge Road
Trenton, NJ 08619 USA
Tel: +1 609.587.9797 Fax: +1 609.587.1970
Email: info@princetoninstruments.com

INDUCTIVELY COUPLED PLASMA MASS SPECTROMETER (ICP-MS)

As ICP-MS has evolved to become the premier technique for trace metals analysis, Agilent has been at the forefront of development and design, introducing many important innovations and achieving unrivalled sales success. With the 7700 Series ICP-MS, Agilent continues to revolutionize the ICP-MS landscape by increasing performance, reducing interferences and improving productivity – all while making the technology easier to use, maintain and service.

System Specifications of ICP-MS 7700 (7700 Series ICP-MS, Agilent Technologies, Inc., USA)

This rugged new instrument incorporates several new design features including a simple-to-use Octopole Reaction System (ORS) for interference removal in high matrix samples. Redesigned ion optics and a novel, high transmission reaction cell give the 7500ce 5x sensitivity increase over its predecessor, the 7500c, enabling low ppt detection limits in a wide range of complex matrices. Plasma and cell gases are controlled by a new, Agilent-designed Active Flow Control system, and ICP RF power is supplied by Agilent's new digital-drive RF generator which delivers the highest available coupling efficiency. The 7500ce offers a no-compromise solution to the most difficult analytical challenges and is ready configured to meet strict environmental regulatory requirements including US EPA methods. The 7500ce's unmatched performance features include:

- Elimination of multiple polyatomic interferences on critical elements such as arsenic, selenium, chromium, vanadium and iron, without the need for highly reactive gases or expensive pre-mixed gas blends
- Interference removal without matrix matching, element specific optimization or interference equations
- High sensitivity and low oxides for the best performance even in the most difficult matrices
- Dynamic range from low-ppt for mercury, arsenic and selenium to 1000's of ppm for sodium in the same analytical run, thereby eliminating the need for additional analytical techniques, such as ICP-OES, GFAAS, cold vapor and atomic fluorescence

Ref: http://www.chem.agilent.com/Library/technicaloverviews/Public/5989_0735EN.pdf

Vendor (ICP-MS)

Agilent Technologies Inc.
Santa Clara, California
Oswego, Illinois,
USA
Ph: 1-800-227-9770
Email: agilent_inquiries@agilent.com

KEITHLEY 6517A ELECTROMETER

Chemistry analyzer is used for determining thermostabile proteins.

System Specifications of B-200 Clinical Chemistry Analyzer, Mindray, China

DC Current   200 pA - 20 mA
DC Voltage Range   2 V - 200 V
Frequency Range   0 Hz - 60 Hz
Number of Digits   5.5
Resistance Range   2 Mohm - 200 Tohm

Ref: http://www.ipe.cuhk.edu.hk/Equipment_list/Electrometer.htm
http://www3.imperial.ac.uk/pls/portallive/docs/1/7293199.PDF

Vendor (Electrometer)

Keithley Instruments, Inc.
Cleveland, Ohio,
U.S.A.
Ph: (216) 248-0400; Fax: (216) 248-6168
Email: info@keithley.com

MATRIX ASSISTED LASER DESORPTION IONIZATION TIME OF FLIGHT MASS SPECTROMETER (MALDI-TOF AND MALDI-TOF/TOF)

Matrix-assisted laser desorption/ionization (MALDI) is a soft ionization technique used in mass spectrometry, allowing the analysis of biomolecules (biopolymers such as DNA, proteins, peptides and sugars) and large organic molecules (such as polymers, dendrimers and other macromolecules), which tend to be fragile and fragment when ionized by more conventional ionization methods.

System Specifications of ultrafleXtreme™ MALDI-TOF by Brunker Corporation

UltrafleXtreme™ is the most advanced MALDI-TOF/TOF mass spectrometer. Combining true 1kHz speed in both TOF and TOF/TOF modes with ultrahigh performance and extreme flexibility for a broad variety of complementary research, clinical and applied proteomics applications.

- The innovative smartbeam-II™ laser enables ultra-high data acquisition speed in both MS and MS/MS at full systems performance. The well-established proprietary smartbeam laser provides unprecedented analytical and matrix flexibility in workflows from protein tissue imaging, intact proteins analysis, biologics or oligo QC, to LC-MALDI proteomics - all fully enabled at 1-1000 Hz repetition rates.

- Bruker's patented smartbeam technology is already widely accepted as the most viable MALDI imaging laser technology. The ultrafleXtreme now enables laser focus diameters down to 10 μm for high spatial resolution imaging without pixel overlap. Importantly, outstanding spectral quality and signal intensity are maintained at even the smallest laser beam diameters.

- Broadband mass resolving power up to 40,000 enables precision proteomics via Bruker's unique PAN™ technology for highest mass resolution across a very broad mass range, not just at a selected optimum.

- The novel FlashDetector™ combined with a new 4 GHz digitizer and latest advances in electronics provide unmatched mass resolving power up to 40,000 and 1 ppm mass accuracy for highest confidence.

- The novel and unique laser-irradiation self-cleaning MALDI Perpetual™ Ion Source ensures robust, long-term highest-performance operation. Very long MALDI laser lifetime in combination with automated source cleaning in just minutes leads to high uptime and low maintenance costs.

- Latest TOF/TOF technology: The high efficiency and sensitivity of the LID-LIFT process delivers MS/MS spectra with nominal mass resolution for peptides. Typically, full MS/MS data sets can be acquired from low fmol levels within seconds.

***Microflex series by Brunker Corporation, USA is also available for benchtop experiments.

Specifications of MALDI Biotyper Software by Brunker Corporation

The MALDI Biotyper enables an unbiased identification of microorganisms. It can be applied to gram-positive and gram-negative bacteria, yeast and multicellular fungi without any presumptions or pretesting. In a MALDI-TOF mass spectrometer, proteins and peptides are separated by their mass – revealing a characteristic peak pattern, the individual molecular fingerprint. A unique and continuously enlarged library containing thousands of reference entries enables secure identification of your current sample with respect to the references using sophisticated pattern matching algorithms. The MALDI Biotyper covers applications from clinical microbiology, food and feed safety and analysis, as well as industrial quality control.

Ref: http://www.bdal.com/products/maldi/ultraflextreme-series/details.html
http://www.bdal.com/solutions/clinical/microorganism-id/details.html

Vendor (MALDI-TOF AND MALDI-TOF/TOF)

Bruker Corporation
3500 West Warren Avenue
Fremont, CA 94538
USA
Ph: +1 (510) 683-4300; Fax: +1 (510) 687-1217
Email: ms-sales@bdal.com

MICRO-CENTRIFUGE

A laboratory centrifuge is a piece of laboratory equipment, driven by a motor, which spins liquid samples at high speed. There are various types of centrifuges, depending on the size and the sample capacity. Micro-centrifuge device contains small tubes from 0.2 ml to 2.0 ml (micro tubes), up to 96 well-plates, compact design and small footprint; up to 30.000 g.

Ref: http://en.wikipedia.org/wiki/Laboratory_centrifuge

System Specifications of Centrifuse 5430, Eppendorf AG, Germany

Max. rcf:	30, 130 g
Max. rpm:　17,500 1/min	
Max. capacity:	30 x 1.5 / 2.0 ml
No. of rotors:	8
Acceleration time to max. speed:	<25 s
Braking time from max. speed:	<25 s
Soft ramp:　adjustable	
Noise level: <67 dB(A)	
Dimensions in cm (W x D x H):	33 x 42 x 25
Weight without rotor:	29 kg
Power supply:	230 V/50–60 Hz
Power requirement:	475 W

Ref: http://www.eppendorf.com/int/img/landingpages/Centrifuge_5430_en.pdf

Vendor (MICRO-CENTRIFUGE)

Eppendorf AG
Barkhausenweg 1, 22339 Hamburg
Germany
Phone: +4940538010
Email: info@eppendorf.com

Thermo Fisher Scientific Inc
1400 Northpoint Parkway,
Suite 50 West Palm Beach,FL 33407, USA
Ph: +1 800 532 4752
Email: info@thermoscientific.com

MICRO SYRINGE PIPETTE

System Specifications of 701N MICROSYRINGE PIPET, Hamilton, Switzerland

Microsyringe Pipettes have a special 3/4" electrotapered needle for easy attachment of PTFE tips. Your sample only wets the disposable PTFE tips, eliminating sample cross-contamination. The spring-loaded plunger and adjustable stops ensure excellent reproducibility. An adjustable stop allows the transfer of samples from .2 to 3 µL.

Ref: http://www3.imperial.ac.uk/pls/portallive/docs/1/7293199.PDF

Vendor (Micro Syringe Pipette)

Hamilton Bonaduz AG
Via Crusch 8
CH-7402 Bonaduz/Switzerland
Tel: +41-(0)81-660-60-60; Fax: +41-(0)81-660-60-70
E-mail: contact@hamilton.ch

MULTI-CHANNEL ELECTROCHEMICAL DETECTOR

Multi-array electrochemical detection produces qualitative information for compound identification, resolves coeluting peaks, determines peak purity; simplifies sample prep, and measures multiple analytes per sample.

Ref: http://www.dionex.com/en-us/products/liquid-chromatography/lc-modules/detectors/electrochemical/coularray/lp-85243.html

System Specifications of 5600A CoulArray Multi-Channel Electrochemical Detector,

Number of Electrodes	Choice of 4, 8, 12, or 16 coulometric electrodes
Potential Range	Independent potential control for each electrode from -1V to +2V in 1mV increments
Current Ranges	$\pm$ 50 nA, $\pm$ 5 μA, $\pm$ 100 μA autoranged; full scale for each electrode, displayed from 10 pA to 100 μA
Acquisition Rate	Selectable: 2 or 10 Hz
Autozero	Up to 6 μA on the 50 μA and 5 μA scale and 100 μA on the 100 μA scale
Output Noise	< 5 pA peak to peak (10 MΩ, 2 μF, low filter setting)
Output Resolution	30 fA on 50 nA gain range: 3 pA on 5 μA gain range, and 47 pA on 100 μA gain range
Power	100-120 V, 60 Hz; 230-240 V, 50 Hz; 36 VA
External Start Inputs	2
Operating Temperature Range (Instrument)	-10° C to 35° C
Operating Temperature Range (Cells)	-10° C to 45° C
Dimensions and Weight: Detector	44.5 × 26 × 46.2 cm (17.5 × 10.25 × 18.3 in); 12 kg (26 lbs)
Dimensions and Weight: Organizer	44.5 × 26 × 44.2 cm (17.5 × 10.25 × 17.4 in); 6.8 kg (15 lbs)
Certifications	UL, CSA, CE

Ref: http://www.dionex.com/en-us/products/liquid-chromatography/lc-modules/detectors/electrochemical/coularray/lp-85243.html

Vendor (Multi-Channel Electrochemical Detector)

Dionex s
1228 Titan Way
P.O. Box 3603, Sunnyvale, CA
94088-3603, United States
Phone: 14087370700; Fax: +1408730 9403
Email: OrderAdministration@dionex.com

MULTI-CHANNEL POTENTIOSTAT/GALVANOSTAT/IMPEDANCE ANALYZER

System Specifications of VersaSTAT MC, Princeton Applied Research, USA

The VersaSTAT MC is our multi-channel potentiostat/galvanostat/im pedance analyzer, combining fifty years of Princeton Applied Research knowledge and expertise in the development of world leading electrochemical test products with advanced performance from the very latest measurement technology. This, together with easy to use, yet powerful PC software makes the VersaSTAT MC an impressive combina tion of value, performance, and productivity.

Data Acquisition	3 x 16 bit 500k samples per second ADCs synchronized voltage/current/auxiliary
Time Base Resolution (min)	10Âµs (100k samples/second)
Automatic Noise Filters	Enabled/disabled
Voltage Compliance	Â±12V
Current Compliance	Â±650mA (standard) Â±2A (with 2A option)
Potentiostat Bandwidth	1 MHz
Stability Settings	high speed, high stability
Slew Rate	>8V per Âµs typical (no load)
Rise Time (-1.0V to +1.0V)	<350 ns (no load)
Applied Voltage Range	Â±10V
Applied Voltage Resolution	for Â±10mV signal = 300nV for Â±100mV signal = 3ÂµV for Â±1V signal = 30ÂµV for Â±10V signal = 300ÂµV
Applied Voltage Accuracy	Â±0.2% of value Â±2mV
Maximum Scan Rate	5000 Vs^{-1} (50 mV step)
Maximum Scan Range	Â±10V / 300ÂµV
Applied Current Range	Â±full scale (depends on range selected) Â±650mA (standard), Â±2A (with option)
Applied Current Resolution	Â±1/32,000 x full scale
Applied Current Accuracy	Â±0.2% of reading, Â±0.2% of range
Max. Current Range/Resolution	Â±650mA / 60ÂµA
Min. Current Range/Resolution	Â±200nA / 60 pA
Max. Input Range	Â±10V
Bandwidth	10MHz (3dB)
Input Impedance	10 ^{12} Î© in parallel with 5pF (typical)
Leakage Current	5pA at less than 25Â°C
CMRR	60 dB at 100kHz (typical)
Voltage Range	Â±10V
Minimum Resolution	6ÂµV
Voltage Accuracy	Â±0.2% of reading, Â±2mV

Current Ranges	Auto-ranging (8 ranges) 650 mA to 200 nA (8 ranges) 2A to 200nA (with option)
Current Resolution	120 fA
Current Accuracy (DC)	Â±0.2% or reading, Â±0.2% or range
Bandwidth	1MHz (signal â‰¥2mA range typical)
Bandwidth limit filter	Yes
Mode	Potentiostat/Galvanostatic
Frequency Ranges	10ÂµHz to 1MHz
Minimum AC Voltage Amplitude	0.1mV RMS
Sweep	Linear or Logarithmic
Positive Feedback	Yes
Dynamic iR	Yes
Digital inputs/outputs	5 TTL logic outputs, 2 TTL logic inputs
Auxiliary Voltage Input	Meas urement synchronized to V and I Â±10V range, input impedance 10kÎ© Filter: off, 1kHz, 200kHz BNC connector
DAC Voltage Output (Standard)	Â±10V range, output impedance 1kÎ© BNC connector (for stirrers, rotating disk electrode etc.)
Communications Interface	Universal Serial Bus (USB)
Operating System	Windows 7 (64-bit & 32-bit) Windows XP / Vista
PC Specification (minimum)	Pentium 4 (1GHz) / 1GB memory High Data rates may require additional memory
Software	VersaStudio

Ref: http://www.princetonappliedresearch.com/Our-Products/Potentiostats/VersaSTAT-MC.aspx

Vendor (**Multi-Channel Potentiostat/Galvanostat/Impedance Analyzer**)

Princeton Applied Research
801 South Illinois Avenue
Oak Ridge, TN 37830
USA
Ph: 865-425-1289; Fax: 865-481-2410
Email: pari.info@ametek.com

NMR SPECTROMETER

Nuclear magnetic resonance (NMR) is a physical phenomenon in which magnetic nuclei in a magnetic field absorb and re-emit electromagnetic radiation. This energy is at a specific resonance frequency which depends on the strength of the magnetic field and the magnetic properties of the isotope of the atoms.

System Specifications of Varian 600 MHz NMR Magnet (Varia n, Inc., purchased by Agilent Technologies, Inc., USA)

The Varian 600 MHz NMR system consists of a highly homogeneous superconducting magnet (600 MHz 1H, 14.1 Tesla), housed within an ultra low-loss helium cryostat with a nominal room-temperature bore of 54 mm. The magnet features excellent fringe field characteristics, improved magnet shielding from external perturbations and minimized ceiling height. These facilitate ease of system siting and improve operational safety.

NMR frequency:	600 Hz
Drift: 10 Hz/Hr	
Superconducting shim coils	Z1, Z2, Z3, X, Y, ZX, ZY, XY, X2-Y2
Radial 5 G stray field	
from magnet center	97 cm/38.2 in.
Magnet radius (including flanges)	51 cm/20.1 in.
Axial 5 G height above floor	265 cm/104.3 in.
Axial 5 G depth below floor	25 cm/9.8 in.
System weight, operational	1460 kg/3220 lb
Vibration isolation using	
pneumatic legs	Included as standard
Minimum ceiling height	341 cm/134.3 in.
Minimum ceiling height	
(optional transfer)	310 cm/122.0 in.
Liquid helium refill volume	92 L
Liquid helium hold time	120 days
Liquid nitrogen refill volume	143 L
Liquid nitrogen hold time	21 days

System Includes:

- Main magnet housed within a low-loss cryostat
- Set of anti-vibration legs
- Liquid helium level probe and meter
- Liquid nitrogen level probe and meter
- Liquid helium transfer siphon and extension tube
- Braided liquid nitrogen transfer line
- Room temperature shim set
- Helium and nitrogen gas flow meter

Ref: http://www.varianinc.com/image/vimage/docs/products/nmr/magnets/shared/Varian_Magnet_600MHz.pdf

Vendor (**NMR Spectrometer**)

Agilent Technologies Inc.
Santa Clara, California
Oswego, Illinois,
USA
Ph: 1-800-227-9770
Email: agilent_inquiries@@agilent.com

ORIGINPRO (DATA ANALYSIS AND GRAPHING SOFTWARE)

Origin is a software application with tools for data analysis, publication-quality graphing, and programming.

Data Analysis
Origin contains powerful tools for all of your analytic needs, including peak analysis, curve fitting, statistics, and signal processing. To make data analysis more efficient, Origin supports many common formats for importing data, and exporting results. Multi-sheet workbooks and an integrated Project Explorer help you organize your Origin projects. Streamline your workflow by saving workbooks as analysis templates for repeat use.

Graphing
With over 70 built-in graph types, Origin makes it easy to create and customize publication quality graphs to suit your needs. Many popular contour, 2D, and 3D graph types are supported, as are specialty graphs such as windrose, stock (OHLC), ternary (including ternary-contour), 2D vector and 3D vector, and several statistical plots.

OriginPro offers all of the features of Origin plus extended analysis tools in the following areas:

* Peak Fitting
* Surface Fitting
* Statistics
* Signal Processing
* Image Handling

Ref: http://www.originlab.com/index.aspx?go=Products/OriginPro

Vendor (ORIGINPRO)
Price: USD 960 (Education purpose)
OriginLab Corporation
One Roundhouse Plaza, Suite 303
Northampton, MA 01060
USA
Ph: 1-413-586-2013; Fax: 1-413-585-0126

POTENTIOSTAT/GALVANOSTAT ELECTROCHEMICAL SYSTEMS

A potentiostat is the electronic hardware required to control a three electrode cell and run most electroanalytical experiments. A galvanostat is a control and measuring device capable of keeping the current through an electrolytic cell in coulometric titrations constant, disregarding changes in the load itself.

Ref: http://en.wikipedia.org/wiki/Potentiostat

System Specifications of Autolab PGSTAT302N - Hig h performance (Metrohm Autolab B.V, Nethelands)

This high end, high current potentiostat/galvanostat, with a compliance voltage of 30 V and a bandwidth of over 1 MHz, combined with our FRA2 module, is specially designed for electrochemical impedance spectroscopy. The PGSTAT302N is the succesor of the popular PGSTAT30. The maximum current is 2 A, the current range can be extended to 20 A with the BOOSTER20A, the current resolution is 30 fA at a current range of 10 nA.

Electrode connections	2, 3 and 4
Potential range	+/- 10 V
Compliance voltage	+/- 30 V
Maximum current	+/- 2 A (20 A with BOOSTER20A)
Current ranges	1 A to 10 nA (100 pA with ECD module)
Potential accuracy	+/- 0.2 %
Potential resolution	0.3 µV
Current accuracy	+/- 0.2 %
Current resolution	0.0003 % (of current range)
Input impedance	> 1 TOhm
Potentiostat bandwidth	1 MHz
Computer interface	USB
Control software	NOVA

Ref: http://www.ecochemie.nl/Products/Echem/PGSTAT302N.html

VA Measuring Stands

Specifications of 663 VA Stand for Autolab potentiostats

Workstation for Autolab potentiostats. Complete accessory with all electrodes for a complete measurement system: Multi Mode Electrode (MME), Ag/AgCl reference electrode, and glassy carbon (GC) auxiliary electrode. Option: Rotating Disk Electrode (RDE). Without cable and power supply. Connection to Autolab potentiostats with IME663 interface.

Dimensions in mm (W/H/D)	290/295/420
Dimensions remark	Height with fully raised cover: 570 mm

Ref: http://products.metrohm.com/prod-26630020.aspx

VA Trace Analyzer

Specifications of 797 VA Computrace for trace analysis

797 VA Computrace is a modern voltammetric measuring stand that is connected to a PC via a USB port. The PC software provided controls the measurement, records the measuring data and evaluates it. As a result of the well-laid-out program structure the operation is very simple. The built-in potentiostat with galvanostat guarantees the highest degree of sensitivity with reduced noise. Voltammetry system for trace analysis and training. Complete accessories with all electrodes for a complete measuring system: Multi-Mode Electrode (MME), Ag/AgCl reference electrode and platinum auxiliary electrode.

Ref: http://products.metrohm.com/Produktgruppen/polarograph/Kompakt-VA-Ger%C3%A4te/prod-27970010.aspx

VA Autosampler

Specifications of 863 Compact VA Autosampler

Autosampler for the basic automation of VA systems, with integrated peristaltic pump for sample transfer. The sample changer can be used in combination with the 797 VA Computrace. Capacity: 18 samples.

Ref: http://products.metrohm.com/prod-28630020-3.aspx

Vendor (Potentiostat/Galvanostat Electrochemical Systems, VA Trace Analyzer, VA Autosampler, VA Measuring Stands)

Metrohm AG
Ionenstrasse
9101 Herisau
Switzerland
Ph: +41 71 353 8585; Fax: +41 71 353 8901
Email: info@metrohm.com

PTFE Capillary

System Specifications of PTFE Capillary, Metrohm, Switzerland

Internal diameter: 0.25 mm
Lengh: 1 m

Ref: http://products.metrohm.com/accessories/connector-plastic/ptfe-etfe-tubing/prod-61803150.aspx

Vendor (PTFE Capillary)

Metrohm AG
Ionenstrasse, 9101 Herisau
Switzerland
Ph: +41 71 353 8585; Fax: +41 71 353 8901
Email: info@metrohm.com

PYROELECTRIC ENERGY METER (Gentec-EO, USA)

Our pyroelectric energy meters cover a very wide range, going from nanojoules to several tens of joules per pulse.

Key features of QE-25 Energy Detector, Gentec-EO, USA

- Modular Concept Increase the power capability of your detector: 2 different cooling modules
- Low Noise Level 2 µJ for the MB coating

- Test Target Included With the MB models
- Available with Metallic Absorber High Repetition Rate (6000 Hz)
- Smart Interface Containing all the calibration data

System Specifications of QE-25, Energy Dectector, Gentec-EO, USA

Can be downloaded via link **http://gentec-eo.com/Content/uploads/downloads/2-Energy_Detectors/7_QE25/1-Documentation/1-Specifications_Sheet/QE25_2011_V1.02.pdf**

Vendor (Pyroelectric Energy Meter)

Gentec-EO USA Inc.
5825 Jean Road Center
Lake Oswego, OR 97035
USA
Ms Nathalie Becotte
Tel : (418) 651-8003 (ext. 310), Fax : (418) 651-1174
Email: nbecotte@gentec-eo.com

Q-SUN XENON TEST CHAMBERS

System Specifications of Q-Sun Xenon Test Chambers, Q-Lab, USA

All testers include, full-spectrum xenon arc lamps, the solar eye irradiance control system, a choice of irradiance set poits (340 nm, 420 nm or TUV), a choic of filters, AutoCal system that allows quick and easy calibration, a large specimen capacity to acommoate 3D parts. It also includes ASTM, ISO, DiN, SAE, BS, ANSI, AATCC.

Ref: http://www.q-lab.com/EN_Weblit/Q-Sun-LX-5050_4_web.pdf

Vendor (Xenon Test Chambers)

Q-Lab Instruments Division
800 Canterbury Road
Cleveland, OH 44145
USA
Tel.: +1-440-835-8700
E-Mail: Info@q-lab.com

RAMAN SPECTROSCOPY

Raman spectroscopy is a light scattering technique, and can be thought of in its simplest form as a process where a photon of light interacts with a sample to produce scattered radiation of different wavelengths. Raman spectroscopy is extremely information rich, (useful for chemical identification, characterization of molecular structures, effects of bonding, environment and stress on a sample). Historically, the technique of Raman spectroscopy was not that widely taught within university courses, even though the scattering process itself was established as far back as 1928 by Professor C.V Raman. Raman spectroscopy has become an important analytical and research tool. It can be used for applications as wide ranging as pharmaceuticals, forensic science, polymers, thin films, semiconductors and even for the analysis of fullerene structures and carbon nano-materials.

Raman spectroscopy is used to study the vibrational modes of molecules and their functional groups. As such, it is complementary to infrared spectroscopy and can also be used to identify molecules and their functional group in a similar fashion. However, it has certain advantages to infrared spectroscopy which range from the ability to study molecules that are not IR active as well as not being susceptible to water contamination, to name just two.

Raman spectra are acquired by measuring the inelastic scattering of monochromatic photons from a sample. The photons of a monochromatic light source are focused onto a sample and interact with the molecular vibrational modes of the sample. The scattered photons are either shifted either up or down, in terms of their energy. The amount of shift is dependant upon the vibrational mode of the molecule or functional group from which they were scattered. This shift is measured by the Raman spectrometer and results in a spectra of the vibrational modes of the sample.

In modern Raman spectrometers, a laser is used as the monochromatic light source. These instruments also include a monochromator fitted with a sensitive detector that is used to detect the inelastically scattered light. The Raman microspectrometer also includes a microscope so that the Raman spectra of microscopic samples and microscopic areas of larger samples may be measured. In a typical Raman microspectroscopy experiment, the laser is focused through the microscope objective and used to illuminate the sample. The same objective then collects the Raman scattered light which is then focused onto the entrance aperture of the Raman spectrometer which consists of a monochromator and a detector. Modern Raman microspectrometers use charge coupled devices (CCD) due to their sensitivity and stability.

The CCD measures the intensity of Raman scattered light at each wavelength relative to the laser line and the result is displayed on a computer by the instrument control software.

The CRAIC Apollo™ Raman microspectrometer is designed to attach to an optical microscope. It is can also be added to a UV-visible microspectrometer to give the microspectrophotometer Raman spectroscopy capabilities in addition to UV-visible and fluorescence microspectroscopy features. CRAIC Apollo™ Raman microspectrometers offer a range of laser wavelengths from the ultraviolet to the near infrared with user defined spectral resolution and a spectral range that begins at 200 to 2100 cm-1. The CRAIC Apollo™ Raman microspectrometer gives the user to the ability to easily measure the Raman spectrum of microscopic samples.

CRAIC Apollo™ Raman microspectrometer prices begin at US$30000 and offer numerous different laser excitation wavelengths as well as adapters for different microscopes and microspectrophotometers.

Contribution by CRAIC Technologies, USA

System Specifications of small bench-top spectrometer systems (T64000, Horiba Scientific, Japan)

Focal length	640mm, all stages
Aperture[**]	f/7.5
Dispersion[**] Additive triple	0.23 nm/mm
Dispersion[**] Single (direct or subtractive double)	0.7 nm/mm
Gratings	76 x 76 mm^2; selection from over 50 gratings including the patented PAC gratings
Drive mechanism	Sine bar
Step Size[*]	0.00066 nm
Mechanical range[*]	0 to 1000 nm (and above)
Slits	0 to 2 mm wide; 0.5,1.,2.5,5,15 mm high
(subtractive intermediate)	0 to 50 mm wide; 0 to 15 mm high
(spectrograph entrance)	0 to 25 mm wide, 0 to 15 mm high
Spectrograph port	25 - 30 mm clear aperture

[*]Based on 1800 gr.mm gratings, [**]dependent upon wavelength

Spectral Resolution and Coverage

Configuration		Single (direct or double filter stage)	Additive Triple
Grating (gr/mm)	Mechanical Range	CCDCoverage (1")**	CCD Coverage (1")**
300	0-6000 nm	133 nm	38.0 nm
600	0-3000 nm	62 nm	21.0 nm
1200	0-1500 nm	30 nm	9.1 nm
1800	0-1000 nm	17 nm	5.7 nm
2400	0-750 nm	11 nm	4.1 nm
3600	0-500 nm	6 nm	2.0 nm

** 1 nm corresponds to ca.

100 cm-1 at 320 nm, 40 cm-1 at500 nm, 15 cm-1 at 750 nm

Vendors (Raman Spectroscopy)

HORIBA Ltd
Mr Makoto Hashimoto
2 Miyanohigashi, Kisshoin
Minami-Ku 601-8510 KYOTO Japan
Ph: +81 75 313 8121, Fax: +81 75 321 5725
Email: makoto.hashimoto@horiba.com

CRAIC Technologies
948 N. Amelia Ave.
San Dimas, CA 91773
USA
Ph: +1 310 5738180, Fax: +1 310 5738182
Email: sales@microspectra.com

PerkinElmer Inc
940 Winter Street
Waltham, Massachusetts 02451
USA
Tel : 800-762-4000, Fax : 001 203-944-4904
Email: CustomerCareUS@perkinelmer.com

SCANNING ELECTRON MICROSCOPE

A scanning electron microscope (SEM) is a type of electron microscope that images a sample by scanning it with a high-energy beam of electrons in a raster scan pattern. The electrons interact with the atoms that make up the sample producing signals that contain information about the sample's surface topography, composition, and other properties such as electrical conductivity.

Ref: http://en.wikipedia.org/wiki/Scanning_electron_microscope

System Specifications of VEGA3 SBU, SEM, TESCAN, Czech Republic

SEM Column

Resolution (SE)	3 nm at 30 kV
	8 nm at 3 kV
Magnification	1x to 1000000x
Electron Gun	Tungsten heated filament
Resolution - Low Vacuum (BSE, LVSTD)	3.5 nm at 30kV
Accelerating Voltage	200 V to 30 kV
Probe Current	1 pA to 2μA
Chamber	
Internal diameter	160 mm
Door width	120 mm
Number of ports	10
Chamber suspension	mechanic
Working Vacuum	
High vacuum mode	$< 9 \times 10\text{-}3$ Pa
Medium vacuum (SBU)	3 – 150 Pa
Low vacuum (SBU)	3 – 500 Pa
Stage	
Eucentric type	
Movements	X = 45 mm – motorized
	Y = 45 mm – motorized
	Z = 27 mm – manual
	Z" = 6 mm – manual
Eucentrical	– manual
Specimen height	maximum 36 mm
EasyEDX Microanalyser	
Detection range	from B(5) to Am(95)
Max. input count rate	150 kcps
Detector type	SDD Detector
Energy Resolution	133 eV (Mn Ka) at 100 kcps
Detector cooling	Peltier couple (LN2 fr)

Ref: http://www.tescan.com/product.php?id_menu=35&id=32&name=VEGA+3+SB

Vendor (Scanning Electron Microscope)

TESCAN, a.s.
Libusina tr. 21
623 00 Brno
Czech Republic
Tel: +420 547 130 411; Fax: +420 547 130 415
Email: info@tescan.cz

SIMUL 5.0

Simul is a freeware program which performs simulation of the movement of ions in liquid solutions in the electric field. It solves sets of nonlinear partial differential equations and nonlinear algebraic equations describing the continuity of ionic movement and acid-base equilibria. Simul is useful for people engaged in electromigration separation methods, e.g. capillary electrophoresis and isotachophoresis.

Software download Link: http://web.natur.cuni.cz/gas/Simul50.exe

Ref: http://web.natur.cuni.cz/gas/

Vendor (Simul 5.0)

Villa Labeco spol. s. r.o.
Chrapčiakova 1. 052 01 Spišská Nová Ves
Slovakia
Tel: 00421 53 44 260 32; Fax: 00421 53 4196
104
Email: villa@spisnet.sk

THERMAL CYCLER & RT-PCR

System Specifications of C1000 Touch Thermal Cycler, Bio-Rad, USA

Input power	850 W, maximum
Frequency 50–60 Hz, single phase	
Display	8.5 in. (21.6 cm) LCD display and touch screen
Ports 5 USB A, 1 USB B	
Fuses	Two 6.3 A, 250 V, 5 x 20 mm
Memory	>1,000 typical programs onboard, unlimited with USB flash drive expansion
Dimensions (W x D x H)	33 x 46 x 20 cm (13 x 18 x 8")
Weight 10 kg (23 lb)	
Temperature control modes	Calculated and block
PCR license Yes	
Programming options	Step-based graphical, automatic
Security features	Optional log-in required mode for regulated environments
Reporting	Exportable run logs, system error logs
Onboard software	Windows CE 6.0
PC compatibility	Windows XP or higher
USB peripheral compatibility	Mouse, USB flash drive
Real-time PCR upgrade	6-channel CFX96 optical reaction module and 5-channel CFX384 optical reaction module
Instant incubation Yes	

System Specifications for CFX96 Touch Real-Time PCR Detection System, Bio-Rad, USA

Base thermal cycler	C1000 Touch
Max ramp rate	5°C/sec
Ave ramp rate	3.3°C/sec
Heating and cooling method	Peltier
Temperature Range	0–100°C
Accuracy ±0.2°C	programmed target at 90°C
Uniformity ±0.4°C well-to-well	within 10 sec of arrival at 90°C
Gradient Yes	
Programmable span	1–24°C

Optical Detection:

Excitation 6 filtered LEDs	
Range of excitation/emission wavelengths 450–730 nm	
Sensitivity	Detects 1 copy of target sequence in human genomic DNA
Dynamic range	10 orders of magnitude
Scan time	Single channel
Fast scan 3 sec	
All channels	12 sec

System

Licensed for real-time PCR	Yes
Wells 96	
Sample size	1–50 µl (10–25 µl recommended)
Electrical approvals	IEC, CE
Dimensions (W x D x H)	33 x 46 x 36 cm (13 x 18 x 14")

Ref: http://www3.bio-rad.com/

Vendor (Thermal Cycler)

Bio-Rad
2000 Alfred Nobel Drive
Hercules, CA 94547
Telephone: 510-741-1000
E-Mail: lsg.orders.us@bio-rad.com

UP MACRO LARGE BEAM LASER ABLATION SYSTEM

The new UP MACRO, a large beam, 266nm UV Nd:YAG laser ablati on system is designed to boost ICP-OES and ICP-MS sensitivity well beyond the abilities of all other 266 laser ablation systems. The UP MACRO produces flat craters ranging from 30μm to 750μm in diameter through precision, aperture imaging. This system also features VERTM, (Vernier Extended Range Increments) functionality that ex tends these aperture imaged flat spot sizes from 20μm to greater than 1100μm.

Flat, pre-calibrated spots from 30μm spots and up to > 750μm.
VERITM functionality for spots from 20μm to >1200μm. ICP-OES detection limits of < 1μg/g (1ppm), are possible.
ICP-MS detection limits of < 1ng/g (1ppb), are possible.
>15J/cm2 energy density. Enough to efficiently ablate glass, steel, noble metals, plastics, ceramics and biological matrices.

Applications for UP Macro Laser Ablation includes bulk and inclusion analysis of metals, forensic analysis of plastics, ceramics, paints, glass, biological tissues, tree rings and gelsSulfate, Sulfite analysis, environmental air filters, wear metals, soils analysis.

UP MACRO Large Beam Laser Ablation System (ESI World, Portland)

Site Requirements

Temperature — 70° ±10° F (21° ±5°C)
Relative Humidity — 20 – 80% non-condensing
Voltage — 100 - 240 VAC, 50/60 Hz
Power — 1000 Watts

Physical Parameters

Length — 25" / 64 mm
Width — 18" / 46 mm
Height — 22" / 56 mm
Weight — 130 lb. / 59 kg

Ref: http://www.esi.com/Products/InterconnectMicromachining/LaserAblation/UPMACRO.aspx

Vendor (Up Macro Large Beam Laser Ablation System)

ESI World
Mike Colucci, Sales Manager,
13900 NW Science Park Drive
Portland, OR 97229-5497
Ph: 503.641.4141, Fax: 503.671.5551
Email: mcolucci@esi.com

UV/VISIBLE SPECTROMETER

Ultraviolet-visible spectroscopy or ultraviolet-visible spectrophotometry (UV-Vis or UV/Vis) refers to absorption spectroscopy or reflectance spectroscopy in the ultraviolet-visible spectral region. This means it uses light in the visible and adjacent (near-UV and near-infrared (NIR)) ranges. The absorption or reflectance in the visible range directly affects the perceived color of the chemicals involved. In this region of the electromagnetic spectrum, molecules undergo electronic transitions. Spectrophotometers were originally developed for the absorption measurement of liquid samples. In recent years, however, high-precision, high-energy spectrophotometers have hit the market due to the rapid increase in reflection and absorption measurements on solid samples, including semiconductors, films, glass, and absorbing materials. In the meantime, the field of life sciences demands high-throughput analyzers offering high-sensitivity and high-speed measurement of ultra-trace samples. The fields of pharmaceuticals and drug manufacture demand security, audit-trail and other GXP- and FDA 21 CFR Part 11-compatibility functions. Therefore, the spectrophotometer software must also support these functions.

System Specifications of UV-3600 UV-VIS-NIR Spectro photometer (Shimadzu Scientific Instruments, USA)

This rugged new instrument incorporates several new design features including a simple-to-use Octopole Reaction System (ORS) for interference The UV-3600 UV-VIS-NIR Spectrophotometer as a new additional model in our product lineup of high-end UV-VIS-NIR spectrophotometers. The UV-3600, a successor model to the UV-3150/ 3101, is designed for the measurement of liquid samples, while the SolidSpec-3700/3700DUV is mainly designed for the measurement of solid samples.

Hardware Specifications:

Wavelength range	185nm - 3300nm
Spectral Bandwidth	8 steps in ultraviolet/visible region 0.1, 0.2, 0.5, 1, 2, 3, 5, 8 nm
10 steps in near-infrared region	0.2, 0.5, 1, 2, 3, 5, 8, 12, 20, 32 nm
Resolution	0.1 nm
Sampling Pitch	0.01 - 5 nm
Wavelength accuracy	UV/VIS region: ±0.2 nm NIR region: ±0. 8 nm
Wavelength repeatability	UV/VIS region: less than ±0.08 nm NIR region: less than ±0.32 nm
Stray light	< 0.00008% (220 nm, NaI)
	< 0.00005% (340 nm, NaNO2)
	< 0.0005% (1420 nm, H2O)
	< 0.005% (2365 nm, CHCl3)
Photometric range	-6 to 6 Abs
Noise	0.00005 Abs or less (500 nm), 0.00008 Abs or less (900 nm), 0.00003 Abs or less (1500 nm)
	Slit width 2nm, RMS value at 1 sec. response
Photometric System	Double beam
Dimensions	1020(W) x 660(D) x 275(H) mm

Software Specifications

UVProbe software for the UV-3600 includes functions for spectrum measurement, quantitation, kinetics and a report generator. Multiple security and audit-trail functions ensure the reliability of data.

Ref: http://www.shimadzu.com/an/spectro/uv/uv3600.html

Vendor (UV/Visible Spectrometer)

Shimadzu Scientific Instruments
7102 Riverwood Drive
Columbia, MD 21046
USA
Phone:1-800-477-1227, Fax: 410-381-1222
http://www.shimadzu.com/contact/index.html

VARIABLE-WAVELENGTH UV-VIS HPLC DETECTORS

The Variable Wavelength Detectors (VWD) are the detector of choice for routine chromatography applications. It offers high sensitivity, low noise and drift, and a wavelength range of 190–900 nm. The VWD is available in two variants. The standard version with a 20 Hz data rate is economically affordable for laboratories on a limited budget. The rapid separation version offers a data rate of up to 100 Hz to perfectly detect even the narrowest peaks. Both detectors are designed to meet the demanding requirements of today's laboratories.

System Specifications of Variable Wavelength Detector, Dionex, USA

Data Collection Rate	Single wavelength up to 100 Hz (VWD-3100)
	Single wavelength up to 200 Hz (VWD-3400RS, under Chromeleon® 7.1 software control)
	Multiple wavelength up to 5 Hz (VWD-3400RS)
Noise, Single Wavelength < ±3.5	µAU (typical < ±2.5 µAU) at 254 nm, time constant 1 s, dry analytical flow cell
Noise, Multiple Wavelength (VWD-3400RS)	< ±10 µAU (typical < ±7 µAU) at 254 nm and 280 nm, 2 s time constant, dry analytical
Drift < 1 × 10-4 AU/h at 254 nm	and dry analytical flow cell.
Linearity < 5% RSD at 2.	5 AU at 272 nm based on ASTM
Light Source	Deuterium lamp, tungsten lamp (Tungsten lamp optional on VWD-3100)
	Temperature control for both lamps
Wavelength Range	190–900 nm The tungsten lamp is recommended for wavelengths > 600 nm.
Wavelength Accuracy	±1 nm (over detector lifetime)
Wavelength Repeatability ±0.1 nm	
Optical Bandwidth	6 nm at 254 nm
I/O Interfaces Four digita	l inputs, four digital outputs
Analog Outputs	Two analog outputs available as an option via DAC plug-in module
	Software selectable: absorbance, 20-bit resolution, 0–1V (full range) and 0–10 V (full range with adjustable mAU ranges).
Safety Features	Power-up diagnostics of optics, cooling fans, motors, and electronic leak sensor
Power Requirements	85–260 V AC, 50/60 Hz, max. 150 W
Wide range (automatic voltage selection)	

Dimensions (h × w × d)   16 × 42 × 51 cm (6.3 × 16.5 × 20 in.)

Ref: http://www.dionex.com/en-us/products/liquid-chromatography/lc-modules/detectors/rslc-variable-wavelength/lp-72515.html

Vendor (Variable-Wavelength UV-VIS HPLC Detectors)

Dionex
1228 Titan Way
P.O. Box 3603, Sunnyvale, CA
94088-3603, United States
Phone: 14087370700; Fax: +1408730 9403
Email: OrderAdministration@dionex.com

A Guide to Research in Electro Biochemical Technology

Acknowledgements

Selected extended abstracts from the 10th Workshop of Physical Chemists and Electrochemists

Credit for the quality of the conference proceedings goes first and foremost to the authors. They contributed a great deal of effort and creativity to produce this work, and we are very thankful that they chose "10th Workshop of Physical Chemists and Electrochemists" as the place to present it. All of the authors who submitted papers, both accepted and rejected, are responsible for keeping the papers program vital.

Credit also goes to the Paper Committee members, (Dr. Rene Kizek, Dr. Jaromir Hubalek & Dr. Vojtech Adam) who donated enormous blocks of time from busy schedules to carefully read and evaluate all submissions. Each submission received four reviews, two by committee members, one by the appointed reviewers by the editor of Journal of Biochemical Technology and one by the editorial members of Journal of Biochemical Technology. Committee members were chosen for their leadership in the field, their reputation for honesty and good judgment, and their capacity to enjoy and appreciate other people's work. They all scrupulously avoided any involvement with decisions on papers they were connected with in any way. We also like to thank all of co-authors of these proceedings for their helpfulness, patience and, most of all, great collaboration.

We do very believe that reading of this special volume will be interesting for you.

R R Siva Kiran and Libuse Trnkova

10th Anniversary Workshop of Physical Chemists and Electrochemists in Brno, Czech Republic

Two days in June (23th and 24th) 2010 represented the time of the 10th Workshop of Physical Chemists and Electrochemists for scientists who work in this field. Since the first meeting, two of the Brno Universities, Masaryk University and Mendel University in Brno, have sharing the organization of this conference. The conference was not only under patronage the chancellors of both universities (Prof. Ing. Hlušek Jaroslav, CSc., Dr.h.c. and Prof. PhDr. Petr Fiala, Ph.D., LL.M.), but also under patronage of deans of the faculties, Doc. RNDr. Jaromir Leichmann, Ph.D. Faculty of Science of Masaryk University and Prof. Ing. Ladislav Zeman, CSc. Faculty of Agronomy of Mendel University in Brno.
a
The first Workshop of Physical Chemists and Electrochemists was held in the spring of 1998. Since then, participation as well as the number of presented contributions, both in the form of posters or lectures, has been gradually increased. This year, the number of contributions exceeded eighty (Book of Proceedings from 10th Workshop of Physical Chemists and Electrochemists, Eds. Trnkova L., Neplechova K. And Kizek R., Brno, Czech Republic, 200, pp. 304, ISBN 978-80-7375-396-2). Our invitation for plenary lectures was accepted by six prominent Czech scientists in the field of physical chemistry and the sciences where the physical chemistry plays an important role (biophysical chemistry, biophysics, bioelectrochemistry, biomedical engineering and bioinformatics), Ivo Provazník, Oldřich Pytela, Jiří Homola, Oldřich Dračka, Vladimír Vetterl and Michael Heyrovský.

The conference was held in Mendel University in Brno, which is the Czech Republic's second largest city with a population of nearly 400,000 people. It lies in the central part of Europe and within its two hundred-kilometre radius there are other important European capitals: Prague, Vienna and Bratislava. The city is surrounded by beautiful mixed forests, which offer many opportunities for tourism and cycling. Brno prides itself on many notable historic sites that show evidence of its rich cultural history. Once established as a settlement of merchants eight centuries ago on the junctions of the rivers Svratka and Svitava, it withstood the pressure of both the Hussite and Swedish besiegement, witnessed Napoleon's military expedition to Slavkov, bore the cruel consequences of the Austrian defeat at the Battle of the Three Emperors, became an industrial centre of the Habsburg monarchy called "the Austrian Manchester" and, in the twentieth century, gained a character of a modern city thanks to the construction of new buildings in the functionalist style. The most important example of modern architecture in Brno is The Tugendhat Villa - a historic site inscribed on the UNESCO's World Cultural Heritage List. Brno is a university town with one Government University, five Public Universities with twenty-seven faculties and six Private Universities. The number of students exceeds 80,000. Moreover, thanks to Operational European Union Programs, which will bring into south-Moravian metropolis of more than 500 millions Euros, Brno becomes a centre of excellence in science, research and innovation in the Central Europe. We actively participate in research centres CEITEC, ICRC and SIX.

The main slogan of the 10th Anniversary Workshop of Physical Chemists and Electrochemists was: Beautiful is a scientific debate to deliver new ideas, cooperation, which implements these ideas, is even more beautiful. Therefore, we believe that the workshop has provided new impetus and new ideas for further scientific research in the field of chemical and biological sciences. We hope that besides scientific part the workshop was also socially enjoyable meeting and we look forward to seeing you in Brno 2011. Finally, we wish to thank to all of co-authors of these proceedings for their helpfulness, patience and, most of all, great collaboration and to the editorial staff of Sevas Educational Society for their help and support. Moreover, we would like to acknowledged financial support from the following projects: CZ.1.07/2.3/00/09.0224, NANOSEMED GAAV KAN2081130801 and NANIMEL GACR 102/08/1546, INCHEMBIOL (MSM0021622412) and BIO-ANAL-MED (LC06035) from the Ministry of Education, Youth and Sports of the Czech Republic.

Libuše Trnková

Department of Chemistry,
Faculty of Science,
Masaryk University,
Kotlarska, Brno, Czech Republic